Springer Theses

Recognizing Outstanding Ph.D. Research

Aims and Scope

The series "Springer Theses" brings together a selection of the very best Ph.D. theses from around the world and across the physical sciences. Nominated and endorsed by two recognized specialists, each published volume has been selected for its scientific excellence and the high impact of its contents for the pertinent field of research. For greater accessibility to non-specialists, the published versions include an extended introduction, as well as a foreword by the student's supervisor explaining the special relevance of the work for the field. As a whole, the series will provide a valuable resource both for newcomers to the research fields described, and for other scientists seeking detailed background information on special questions. Finally, it provides an accredited documentation of the valuable contributions made by today's younger generation of scientists.

Theses are accepted into the series by invited nomination only and must fulfill all of the following criteria

- They must be written in good English.
- The topic should fall within the confines of Chemistry, Physics, Earth Sciences, Engineering and related interdisciplinary fields such as Materials, Nanoscience, Chemical Engineering, Complex Systems and Biophysics.
- The work reported in the thesis must represent a significant scientific advance.
- If the thesis includes previously published material, permission to reproduce this must be gained from the respective copyright holder.
- They must have been examined and passed during the 12 months prior to nomination.
- Each thesis should include a foreword by the supervisor outlining the significance of its content.
- The theses should have a clearly defined structure including an introduction accessible to scientists not expert in that particular field.

More information about this series at http://www.springer.com/series/8790

Bjorn Scholz

First Observation of Coherent Elastic Neutrino-Nucleus Scattering

Doctoral Thesis accepted by University of Chicago,
Illinois, USA

 Springer

Bjorn Scholz
University of Chicago
Chicago, IL, USA

ISSN 2190-5053   ISSN 2190-5061 (electronic)
Springer Theses
ISBN 978-3-030-07629-0   ISBN 978-3-319-99747-6 (eBook)
https://doi.org/10.1007/978-3-319-99747-6

This Springer imprint is published by the registered company Springer Nature Switzerland AG
The registered company address is: Gewerbestrasse 11, 6330 Cham, Switzerland

Supervisor's Foreword

The title of Bjorn Scholz's Ph.D. dissertation is rather explicit about its contents: "First Observation of Coherent Elastic Neutrino-Nucleus Scattering." In this work, he describes the experimental effort that led to a finally successful measurement of this long-sought process of neutrino interaction (CEνNS, for short). The experiment was performed at the Spallation Neutron Source facility, sited at Oak Ridge National Laboratory, in Tennessee.

Of all known particles, neutrinos distinguish themselves for being the hardest to detect, typically requiring very large multi-ton devices to secure a tiny probability of interaction. John Updike elegantly described this characteristic neutrino property in his poem "Cosmic Gall":

> The earth is just a silly ball
> To them, through which they simply pass,
> Like dustmaids down a drafty hall

The process first measured in Bjorn's dissertation was conceived as a possibility 44 years ago. It involves the arduous detection of very weak, low-energy signals arising from nuclear recoils (tiny neutrino-induced "kicks" to atomic nuclei). The difficulties involved in its detection were foreseen by Daniel Freedman, when he wrote in his seminal theory paper: "*Our suggestion may be an act of hubris, because the inevitable constraints of interaction rate, resolution and background pose grave experimental difficulties.*"

Nevertheless, CEνNS leads to a much larger probability of neutrino interaction when compared to all other known mechanisms. As a result of this, miniaturized neutrino detectors (Bjorn's was handheld at 14 kg) now suddenly become a reality. This process, which a large community of researchers plans to continue studying, will also provide the opportunity to study fundamental neutrino properties presently beyond the sensitivity of other methods. During the decades-long interim between theoretical proposal and experimental observation, phenomenologists elaborated on a multitude of CEνNS applications: searches for nonstandard neutrino interactions, their electromagnetic properties, studies of nuclear structure and of some dark matter models, etc. In this sense, Bjorn's dissertation can accurately be described as the departure point for a new branch in experimental neutrino physics.

Perhaps more appealing to the popular imagination are the "neutrino technologies" that can now be envisioned through the use of miniaturized detectors. At the time of this writing, a number of international groups embark on realizing precisely this. A first possibility is to use compact CEνNS-based detectors to monitor nuclear reactors against the diversion of weapons-grade fuel material, by providing a precise measurement of the (anti)neutrino flux stemming from a power reactor core. Within our University of Chicago group we are developing specialized germanium detectors capable of this. While this is not reflected in his dissertation, during his time as a graduate student Bjorn also made significant contributions to our understanding of the response of germanium to CEνNS-induced nuclear recoils, facilitating this next step.

Bjorn's dissertation represents the culmination of several decades of work in the field of low-background radiation detection techniques. Many of these have been developed with the goal of catching hypothetical weakly interacting dark matter particles in the act of interaction, a thus far fruitless endeavor. By applying them to the detection of neutrinos, Bjorn's work has provided a solid return on this investment.

Chicago, IL, USA   Juan I. Collar
August 2018

Acknowledgments

First and foremost I would like to thank my advisor, Juan Collar. Juan has given me a nearly endless number of opportunities to learn new things and to grow as a scientist. He gave me ample space and time to work on my own, but also provided guidance whenever I needed it. He supported me and my research with a passion I have rarely seen. His physics (and Jazz) knowledge is unrivaled. In short, I could have never wished for a better advisor.

Next I have to deeply thank Philipp Barbeau and Grayson Rich for all the help and input they provided for my analysis. Phil has never failed to miss even subtle flaws in my reasoning, but also always provided ample input on how to address potential issues. I will eternally be grateful for Grayson's help during the quenching factor measurements at TUNL. Without his help I would have never made it through the misery of that measurement.

I also want to thank Nicole Fields for all the work she has put into the characterization of the CsI detector. Without her work this thesis would have been impossible. Everyone else from the COHERENT collaboration also deserves my deepest gratitude. Especially Jason Newby and Yuri Efremenko, for all the help they provided at the SNS. Whenever an issue with the CsI detector arose, they promptly fixed it. Without their help many hours of good beam time would have been lost. As a result the significance of the observation presented in this thesis would have been much lower. I also want to thank Alexey Konovalov for all the work he has put into the parallel analysis pipeline of the CEνNS search data. Having two independent analyses come to the same conclusion puts credibility into both.

I am extremely grateful to everyone from AMCRYS-H. Despite the ongoing Ukraine crisis they were able to grow a perfect CsI[Na] crystal that met all the specs we required. I am also eternally grateful to everyone involved in the transportation of the crystal from Kharkiv to the United States. A safe passage to Chicago was far from guaranteed, due to the close proximity of Kharkiv to the contested borderlands.

Finally I'd like to thank my committee Paolo Privitera, Carlos Wagner, and Philippe Guyot-Sionnest for their advice.

Contents

Parts of this thesis have been published in the following journal articles:

1. *Observation of coherent elastic neutrino-nucleus scattering*, D. Akimov, *et al.*, 2017, Science, 357 (6356), pp. 1123–1126
2. *Coherent neutrino-nucleus scattering detection with a CsI[Na] scintillator at the SNS spallation source*, J.I. Collar, *et al.*, 2015, NIM A, 773, pp. 56–65

List of Abbreviations

CCD	charge coupled device
CEνNS	coherent elastic neutrino-nucleus scattering
CFD	constant fraction discriminator
DSNB	diffuse supernova neutrino background
FOM	figure of merit
FWHM	full width at half maximum
HDPE	high-density polyethylene
LAr	liquid argon
MC	Monte Carlo
MOTS	Mercury Off-Gas Treatment System
NI	National Instruments
NIN	neutrino-induced neutron
ORNL	Oak Ridge National Laboratory
PDF	probability density function
PE	photoelectron
PMT	photomultiplier tube
PNNL	Pacific Northwest National Laboratory
POT	protons-on-target
PS	Phillips Scientific
PSD	pulse shape discrimination
PT	pretrace
QE	quantum efficiency
ROI	region of interest
SBA	super-bialkali
SCA	single-channel analyzer
SM	Standard Model
SNO	Sudbury Neutrino Observatory

SNS	Spallation Neutron Source
SPE	single photoelectron
SSA	shielded source area
ToF	time-of-flight
TUNL	Triangle Universities Nuclear Laboratory
WIMP	Weakly Interacting Massive Particle

Chapter 1
Introduction

In the early 1900s physicists had discovered radioactive decay in experiments, but were unable to provide a theoretical description of the process. In particular beta decay, where an electron is emitted from an atom was puzzling, since it seemingly violated both energy and momentum conservation—core principles of physics. To solve this conundrum, Wolfgang Pauli postulated a mysterious particle, the neutrino. The neutrino would carry the missing momentum and energy and allow the process to obey both conservation laws. Based on this then speculative idea, Enrico Fermi developed the first theoretical description of a weak interaction in 1934 [1].

Fermi devised his theory using analogies from the description of electromagnetism, where the interaction between fermions is mediated by the exchange of a photon. He postulated the existence of a vector boson that transmits the weak force the same way the photon transmits the electromagnetic force. However, given the finite and small range of the weak force, and the charge change in beta decay, the particle had to be massive and charged. Fermi's theory however included one major flaw, i.e., for large energies it predicted cross-sections that would violate unitarity [2]. Two major developments in the mid-century helped to replace Fermi's theory with a new description of the weak interaction. First Yang and Mills developed a theory of massless interacting vector bosons in 1954 [3] which would eliminate the violation of unitarity. Second, in 1964 Higgs showed how a theory initially containing only a massless photon and two scalar particles could be transformed into one that contains a massive vector particle and a single scalar [4]. In the late 1960s, Glashow, Weinberg, and Salam used these advances to develop a new theory of the electroweak interaction that is renormalizable and which contains massive vector bosons [5, 6]. This theory was so successful that it became part of the Standard Model (SM). In this model the weak interaction is mediated by three vector bosons, the W^+, W^-, and the Z^0. As the symbology suggests, the first two carry an electric charge, whereas the last one does not. The prediction of this neutral vector boson implied the existence of weak neutral-current interactions, where no charge is exchanged.

© Springer Nature Switzerland AG 2018
B. Scholz, *First Observation of Coherent Elastic Neutrino-Nucleus Scattering*,
Springer Theses, https://doi.org/10.1007/978-3-319-99747-6_1

The search for evidence of such neutral-current interactions required particles to be accelerated to high energies. In the late 1960s the Proton Synchrotron at CERN and the Alternating Gradient Synchrotron facility at the Brookhaven National Laboratory ranked among the most powerful particle accelerators in the world and it is therefore not surprising that both facilities sought to be the first to discover potential neutral-current interactions. However, neither facility was initially able to detect them. Quite the contrary, first results of the CERN heavy liquid bubble chamber experiment put an upper limit on the ratio of neutral to charged-current couplings at less than 3% [7]. Such a low ratio made a possible detection much more difficult than expected. Even though this result was later revised, it is no surprise that it took scientists almost another decade until they observed a weak neutral current in neutrino–nucleus interactions. In 1973 the first observation was made by the Gargamelle experiment at CERN [8] and later confirmed by the Harvard–Penn–Wisconsin experiment at Fermi National Laboratory [9].

In 1974 Daniel Z. Freedman provided the first theoretical description of coherent elastic neutrino-nucleus scattering (CEνNS) as a SM process [10]. In general, when a neutrino scatters off a nucleus the interaction depends on the non-trivial interplay between the neutrino and the individual nucleons. However, if the neutrino–nucleus momentum transfer is small enough, it does not resolve the internal structure of the nucleus, and as a consequence the neutrino scatters off the nucleus as a whole. This purely quantum mechanical effect gives rise to a coherent enhancement of the scattering cross-section, which scales approximately with the square of the target nucleus' neutron number. The resulting scattering cross-section is therefore several orders of magnitude larger than any other neutrino–nucleus coupling (Fig. 1.1).

It might seem surprising that CEνNS had eluded detection for over four decades given its large cross-section. During this time our understanding of the neutrino has grown dramatically and no less than three Nobel prizes have honored fundamental discoveries surrounding the elusive particle and its interactions. Part of the Nobel prize in 1995 was awarded to F. Reines for the first experimental detection of the neutrino [12]. In 2002 the Royal Swedish Academy of Sciences decided to award one half of the Nobel Prize in Physics jointly to R. Davis Jr. and M. Koshiba for the detection of the cosmic neutrino [13]. Most recently the Nobel committee awarded the 2015 prize to T. Kajita and A.B. McDonald for the discovery of neutrino oscillations, which showed that neutrinos have mass [14]. However, during all these years CEνNS remained unobserved. Being a neutral-current process, the only experimental signature of CEνNS consists of difficult-to-detect nuclear recoils with energies of only a few eV_{nr} to keV_{nr}. In what follows the subscript "nr" emphasizes that the energy quoted is that of a nuclear recoil. In conventional radiation detectors only a small fraction of the total energy carried by such a nuclear recoil is actually converted into a detectable scintillation and ionization signal. The fraction between the detectable and the total energy is usually referred to as quenching factor [15], and is typically on the order of a few to a few 10%. To distinguish the quenched, detectable energy from the total energy of a nuclear recoil a subscript "ee" (electron equivalent) is used for the former throughout this thesis.

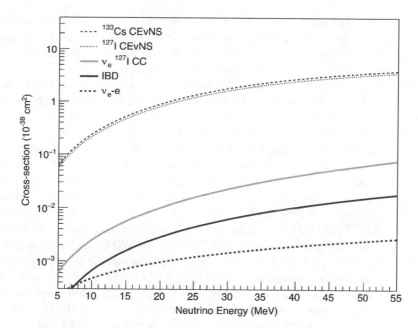

Fig. 1.1 Total cross-section for CEνNS (blue) and other neutrino couplings. Shown are the cross-sections from charged-current (CC) interaction with iodine (green), inverse beta decay (red), and neutrino-electron scattering (dotted red). It is readily visible that CEνNS provides the largest cross-section, dominating over any charged-current interaction for incoming neutrino energies of less than 55 MeV. Plot adapted from [11]

The low energy carried by CEνNS-induced nuclear recoils combined with the small quenching factor had put a potential CEνNS detection out of reach for any conventional neutrino detector like the Sudbury Neutrino Observatory (SNO) or Super-Kamiokande. These detectors are enormous, with detector masses of up to tens of kilo tons. However, they only achieve an energy threshold of a couple of MeV [16], which is much larger than the threshold required to detect CEνNS-induced recoils.

A measurement of the CEνNS cross-section directly tests the description of neutrino–matter interactions as governed by the SM. In addition it also substantiates the coherent cross-section enhancement, which is a crucial component included in the calculation of limits on the Weakly Interacting Massive Particle (WIMP)-nucleus scattering cross-section. For a vanishing momentum transfer between the incoming WIMP and the target nucleus an analogous coherent enhancement of the scattering cross-section is expected. As the neutrino coupling to protons is negligible for CEνNS, the CEνNS cross-section scales with the total neutron number N^2. In contrast, the assumed dark matter coupling to both protons and neutrons is non-negligible and approximately equal [17]. As a consequence, the cross-section for WIMP-nucleus scattering scales with the square of the total nucleon number A^2 instead. Thus the expected enhancement of the cross-section is assumed to be even larger. The absence of such coherent enhancement would shrink the currently explored phase space dramatically. Simultaneously it would also put into question whether we would ever be able to detect a WIMP when its cross-section is several orders of magnitude smaller than what we previously thought.

However, a detection of CEνNS would not only be good news for current dark matter searches by substantiating the coherent cross-section enhancement, it would also require a shift in the modus operandi of the design of dark matter experiments. In the past, the response to an absent WIMP scattering signal usually was to increase the total exposure by building ever bigger detectors. This approach will no longer yield a significant increase in sensitivity, once these dark matter searches run into a new, irreducible background. This background is caused by neutrinos coherently scattering within the dark matter detector. As the neutrinos in question originate from astrophysical sources, such as the sun, or the diffuse supernova neutrino background (DSNB) [18] there is no way to reduce this background component. Since the sole detectable signal for both CEνNS and WIMP scattering is a low-energy nuclear recoil in the detector material , it is impossible to distinguish the two. This inevitably requires a potential WIMP discovery to rely on additional information gained through new approaches, e.g., directional recoil measurements [19]. Figure 1.2 shows the so-called *neutrino floor*, i.e., the hard WIMP discovery limit for future dark matter experiments [18, 25] assuming a SM CEνNS cross-section. It also highlights current best limits on the WIMP-nucleus scattering cross-section by several different dark matter experiments.

Another reason to measure CEνNS stems from the desire to properly understand and model supernovae. When a star undergoes core collapse most of its gravitational energy is converted into energy driving the explosion. Among others, neutrino and nuclear physics are particularly important for describing this process. During the stellar core collapse approximately 99% of the gravitational energy, i.e., $E \approx 3 \times 10^{53}$ erg, is emitted in MeV neutrinos compared to 0.01% as photons [26]. About 0.1 s after the onset of the core collapse the density within the core reaches $\rho_c \sim 10^{12}$ g cm^{-3} [27]. At this point the neutrinos no longer escape the core freely but remain trapped, coherently scattering off the heavy nuclei produced in the core. Given the vast amount of energy released via neutrinos it is thus crucial to have confidence in the CEνNS cross-sections invoked in supernova calculations to properly describe this phase of the collapse.

Given the advances in ultra-sensitive detector technologies over the last decades, which was mainly driven by rare event searches such as dark matter or $0\nu\beta\beta$-decay experiments, it is now feasible to achieve energy thresholds low enough to directly measure CEνNS-induced nuclear recoils and test the SM prediction. There is a large variety of experiments currently being proposed and/or being built to measure CEνNS using different neutrino sources as well as detector technologies.

One group of experiments aims to measure CEνNS using a nuclear reactor as neutrino source [28–31], whereas another group aims to detect CEνNS of neutrinos produced by a stopped pion source [32]. In a stopped pion source a proton beam

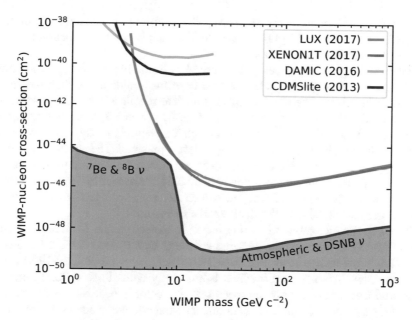

Fig. 1.2 The WIMP discovery limit due to the expected CEνNS of neutrinos from different astrophysical sources is shown in black. Future sensitivity to low-mass WIMPs will be limited by the neutrino emission from the sun, whereas the high-mass sensitivity is limited by atmospheric neutrinos and the DSNB. An increase in exposure would not significantly increase the sensitivity of dark matter experiments beyond this point. Yet there are still multiple ways to improve WIMP sensitivity besides increasing exposure, such as the detection of an annual modulation or directional recoil measurements [19]. Neutrino floor data is taken from [18], CDMSlite [20], LUX [21], XENON1T [22], and DAMIC [23]. Limits from other experiments were omitted for clarity. Additional information on current and planned dark matter experiments is provided in [24]

impinges on a heavy target, which produces mesons. If the target is large enough the produced mesons are stopped and produce several neutrinos in their subsequent decay.

The neutrino flux emitted by a nuclear power plant is several orders of magnitude higher than what is to be expected from even the most intense stopped pion sources. However, the downside is the very low energy carried by these reactor neutrinos. Experiments at a nuclear power plant therefore have to achieve a sub-keV$_{ee}$ energy threshold in order to measure the tiny recoil energies induced in the target material.

A broad array of detector technologies are used in reactor experiments to achieve this sub-keVee energy threshold. Among others there is Ricochet [28], which proposes a 10 kg array of low temperature Ge- or Zn-based bolometers to achieve an energy threshold of 100 eV. In contrast, CONNIE [29] uses an array of charge coupled devices (CCDs) to achieve an effective threshold of only 28 eV. MINER [30] is planning to use cryogenic germanium and silicon detectors in close proximity to a low power research reactor and ν-CLEUS [31] further proposes to use cryogenic CaWO$_4$ and Al$_2$O$_3$ calorimeters to measure CEνNS. All of these experiments build on the expertise gathered from previous efforts to measure CEνNS, such as CoGeNT [33] and should in principle be able to measure CEνNS if all design specifications are met.

The advantage of a stopped pion source lies in the higher energy of neutrinos produced, which is approximately one order of magnitude larger than the energy of reactor neutrinos. The CEνNS-induced nuclear recoils at a stopped pion source therefore carry energies of up to tens of keV$_{nr}$. Even after accounting for the quenching factor this corresponds to detectable energies of up to a few keV$_{ee}$, removing the requirement of a sub-keV$_{ee}$ energy threshold. The neutrinos produced at a stopped pion source include three different flavors ν_μ, $\bar{\nu}_\mu$, and ν_e, whereas the neutrino emission from a nuclear power plant only consists of $\bar{\nu}_e$. However, CEνNS is insensitive to the neutrino flavor and as such all three flavors produced at a stopped pion source contribute to the nuclear recoil spectrum.

The experiment presented in this thesis is located at the Spallation Neutron Source (SNS), a stopped pion neutrino source at Oak Ridge National Laboratory (ORNL). A low-background, 14.57 kg CsI[Na] detector was deployed 19.3 m away from the SNS mercury target, which isotropically emits three different neutrino flavors with well-known emission spectra. The detector is located in a sub-basement corridor, which provides at least 19 m of continuous shielding against beam-related backgrounds and a modest overburden of 8 m.w.e. which reduces backgrounds induced by cosmic-rays. CsI[Na] provides an ideal target material for a CEνNS search [32]. First, the large neutron number of both cesium and iodine results in a large enhancement of the CEνNS cross-section for both elements. Second, its excellent light yield, i.e., the number of single photoelectron (SPE) produced per unit energy, makes it possible to achieve an energy threshold of $\sim$4.5 keV$_{nr}$. After approximately 2 years of continuous data acquisition this thesis presents the first observation of CEνNS at a $6.7 - \sigma$ confidence level.

The theory of CEνNS and its experimental detection is discussed in Chap. 2. Chapter 3 provides detailed information on the SNS and also provides an overview of the activities of the COHERENT collaboration. Several different background studies were performed prior to the deployment of the CsI[Na] detector and are discussed in Chap. 4. Three different CsI[Na] detector calibrations are covered in Chaps. 6–8. The full analysis of CEνNS search data taken at the SNS as well as the first observation of CEνNS is presented in Chap. 9. Chapter 10 provides a brief summary of the findings presented in this thesis and shortly discusses future efforts by COHERENT aiming to decrease the uncertainty on the CEνNS cross-section, test the N^2 dependence, and to put more stringent limits on non-standard interactions.

References

1. E. Fermi, Versuch einer Theorie der β-Strahlen. I. Z. Phys. **88**(3–4), 161–177 (1934)
2. R. Cahn, G. Goldhaber, *The Experimental Foundations of Particle Physics* (Cambridge University Press, Cambridge, 2009)

3. C.N. Yang, R.L. Mills, Conservation of isotopic spin and isotopic gauge invariance. Phys. Rev. **96**, 191–195 (1954). https://link.aps.org/doi/10.1103/PhysRev.96.191
4. P.W. Higgs, Broken symmetries and the masses of gauge bosons. Phys. Rev. Lett. **13**, 508–509 (1964). https://link.aps.org/doi/10.1103/PhysRevLett.13.508
5. S.L. Glashow, Partial symmetries of weak interactions. Nucl. Phys. **22**, 579–588 (1961)
6. S. Weinberg, A model of leptons. Phys. Rev. Lett. **19**, 1264–1266 (1967)
7. M. Block, H. Burmeister, D. Cundy et al., Neutrino interactions in the CERN heavy liquid bubble chamber. Phys. Lett. **12**(3), 281–285 (1964). http://www.sciencedirect.com/science/article/pii/0031916364911047
8. F. Hasert, S. Kabe, W. Krenz et al., Observation of neutrino-like interactions without muon or electron in the Gargamelle neutrino experiment. Nucl. Phys. B **73**(1), 1–22 (1974)
9. F.W. Bullock, Some results from neutrino experiments at CERN. J. Phys. G: Nucl. Phys. **2**(12), 881 (1976)
10. D.Z. Freedman, Coherent effects of a weak neutral current. Phys. Rev. D **9**(5), 1389 (1974)
11. J.I. Collar, D. Akimov, J.B. Albert et al., Observation of coherent elastic neutrino-nucleus scattering. Science **357**(6356), 1123–1126 (2017). http://science.sciencemag.org/content/357/6356/1123
12. A. Nobel Media, The Nobel Prize in Physics 1995. Online. http://www.nobelprize.org/nobel_prizes/physics/laureates/1995/
13. A. Nobel Media, The Nobel Prize in Physics 2002. Online. http://www.nobelprize.org/nobel_prizes/physics/laureates/2002/
14. A. Nobel Media, The Nobel Prize in Physics 2015. Online. http://www.nobelprize.org/nobel_prizes/physics/laureates/2015/
15. B. Scholz, A. Chavarria, J. Collar et al., Measurement of the low-energy quenching factor in germanium using an ^{88}Y/Be photoneutron source. Phys. Rev. D **94**(12), 122003 (2016)
16. M.B. Avanzini, The ^{8}B solar neutrino analysis in Borexino and simulations of muon interaction products in Borexino and double Chooz. Ph.D. thesis, 2011
17. S.J. Brice, R.L. Cooper, F. DeJongh et al., A method for measuring coherent elastic neutrino-nucleus scattering at a far off-axis high-energy neutrino beam target. Phys. Rev. D **89**, 072004 (2014). http://link.aps.org/doi/10.1103/PhysRevD.89.072004
18. J. Billard, E. Figueroa-Feliciano, L. Strigari, Implication of neutrino backgrounds on the reach of next generation dark matter direct detection experiments. Phys. Rev. D **89**(2), 023524 (2014)
19. C.A.J. O'Hare, A.M. Green, J. Billard et al., Readout strategies for directional dark matter detection beyond the neutrino background. Phys. Rev. D **92**, 063518 (2015). https://link.aps.org/doi/10.1103/PhysRevD.92.063518
20. R. Agnese, A. Anderson, M. Asai et al., Search for low-mass weakly interacting massive particles using voltage-assisted calorimetric ionization detection in the SuperCDMS experiment. Phys. Rev. Lett. **112**(4), 041302 (2014)
21. D. Akerib, S. Alsum, H. Araújo et al., Results from a search for dark matter in the complete LUX exposure. Phys. Rev. Lett. **118**(2), 021303 (2017)
22. E. Aprile, J. Aalbers, F. Agostini et al., First dark matter search results from the XENON1T experiment (2017). Preprint. arXiv:1705.06655
23. A. Aguilar-Arevalo, D. Amidei, X. Bertou et al., Search for low-mass WIMPs in a 0.6 kg day exposure of the DAMIC experiment at SNOLAB. Phys. Rev. D **94**(8), 082006 (2016)
24. J. Liu, X. Chen, X. Ji, Current status of direct dark matter detection experiments. Nat. Phys. **13**(3), 212–216 (2017)
25. J. Billard, F. Mayet, D. Santos, Assessing the discovery potential of directional detection of dark matter. Phys. Rev. D **85**, 035006 (2012). https://link.aps.org/doi/10.1103/PhysRevD.85.035006
26. C. Ott, E. O'Connor, S. Gossan et al., Core-collapse supernovae, neutrinos, and gravitational waves. Nucl. Phys. B: Proc. Suppl. **235**, 381–387 (2013). http://www.sciencedirect.com/science/article/pii/S0920563213001576. The XXV International Conference on Neutrino Physics and Astrophysics

27. H.-T. Janka, Neutrino emission from supernovae, in *Handbook of Supernovae* (Springer, Cham, 2016), pp. 1–30
28. J. Billard, R. Carr, J. Dawson et al., Coherent neutrino scattering with low temperature bolometers at CHOOZ reactor complex (2016). Preprint. arXiv:1612.09035
29. A. Aguilar-Arevalo, X. Bertou, C. Bonifazi et al., The CONNIE experiment. J. Phys.: Conf. Ser. **761**, 012057 (2016)
30. G. Agnolet, W. Baker, D. Barker et al., Background studies for the MINER Coherent Neutrino Scattering reactor experiment. Nucl. Instrum. Methods Phys. Res. Sect. A **853**, 53–60 (2017). http://www.sciencedirect.com/science/article/pii/S0168900217302085
31. R. Strauss, J. Rothe, G. Angloher et al., The ν-cleus experiment: a gram-scale fiducial-volume cryogenic detector for the first detection of coherent neutrino-nucleus scattering (2017). Preprint. arXiv:1704.04320
32. J. Collar, N. Fields, M. Hai et al., Coherent neutrino-nucleus scattering detection with a CsI[Na] scintillator at the SNS spallation source. Nucl. Instrum. Methods Phys. Res. Sect. A **773**, 56–65 (2015). http://www.sciencedirect.com/science/article/pii/S0168900214013254
33. P.S. Barbeau, Neutrino and astroparticle physics with P-type point contact high purity germanium detectors. Ph.D. thesis, 2009

Chapter 2
Coherent Elastic Neutrino–Nucleus Scattering

In the SM, CEνNS is mediated via Z^0 exchange (Fig. 2.1). Being a neutral current process CEνNS is insensitive to the flavor of the incoming neutrino resulting in an identical scattering cross-section for all neutrino types apart from some minor corrections. Following Drukier and Stodolsky [1], the differential, spin-independent CEνNS cross-section assuming a negligible momentum transfer (i.e., $q \equiv |\vec{q}| \to 0$) between neutrino and nucleus can be written as:

$$\frac{d\sigma_0}{d\cos\phi} = \frac{G_f^2}{8\pi} \left[Z \left(4\sin^2\Theta_W - 1 \right) + N \right]^2 E_\nu^2 \left(1 + \cos\phi \right), \qquad (2.1)$$

where G_f is the Fermi coupling constant, Z and N are the number of protons and neutrons in the target nucleus, respectively, $\sin^2\Theta_W$ is the weak mixing angle, E_ν is the incoming neutrino energy, and ϕ is the scattering angle. Any contribution from axial vector currents and any radiative corrections above tree level were neglected in Eq. (2.1). Including these contributions at this point would not lead to a deeper understanding of CEνNS but would rather only distract from the core concepts driving CEνNS.

Since for low-momentum transfers $\sin^2\Theta_W = 0.23867 \pm 0.00016 \approx \frac{1}{4}$ [2], Eq. (2.1) can be simplified by eliminating any contribution from proton coupling in a first-order approximation. It is further beneficial to express Eq. (2.1) in terms of the three momentum transfers q between neutrino and nucleus, which yields

$$\frac{d\sigma_0}{dq^2} = \frac{G_f^2}{8\pi} N^2 \left(1 - \frac{q^2}{q_{max}^2} \right) \quad \text{with} \quad q^2 = 2E_\nu^2 \left(1 - \cos\phi \right). \qquad (2.2)$$

Here, q_{max} denotes the maximum momentum transfer at $\phi = 180°$. As the only visible outcome of a CEνNS event is an energy deposition in the detector in the form of a nuclear recoil, it is advantageous from an experimental point of view to

© Springer Nature Switzerland AG 2018
B. Scholz, *First Observation of Coherent Elastic Neutrino-Nucleus Scattering*,
Springer Theses, https://doi.org/10.1007/978-3-319-99747-6_2

Fig. 2.1 Feynman diagram
for CEνNS. Here, ν_i
describes both neutrinos and
anti-neutrinos of any flavor
and N denotes any nucleus

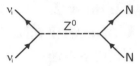

express the differential cross-section in terms of the recoil energy E_r carried by the
target nucleus. It is

$$\frac{d\sigma_0}{dE_r} = \frac{G_f^2}{4\pi} m_a N^2 \left(1 - \frac{m_a E_r}{2E_\nu^2}\right) \quad \text{as} \quad E_r = \frac{q^2}{2m_a}, \tag{2.3}$$

where m_a denotes the nuclear mass of the target material. It is easy to verify that
the maximum induced recoil energy scales with the square of the incoming neutrino
energy, i.e., $E_r^{\max} = 2E_\nu^2/m_a$. As such, it might seem beneficial to choose a neutrino
source with the highest possible energy.

However, Eq. (2.1) only holds in the limit of vanishing momentum transfer
$q \to 0$ and as such $E_r \to 0$. The effective cross-section actually decreases with
increasing q, as the effective de Broglie wavelength $\hbar/q$ approaches the size of the
nucleus. At this point, the target does no longer recoil coherently, but rather scatters
as a collection of individual nucleons. This loss of coherence becomes important
for momentum transfers which satisfy $q R_A \gtrsim 1$, where R_A is the radius of the
nucleus [1] and which is approximately given by $R_A = A^{1/3} \cdot 1.2\,\text{fm}$, where A is
the mass number of the target element [3]. For ^{133}Cs and ^{127}I, this corresponds to a
momentum transfer on the order of $q \approx 30\,\text{MeV}$. Following Lewin and Smith [4],
this effect can be incorporated by introducing a form factor $F(q)$ that only depends
on the momentum transfer and is independent of the nature of the interaction. It
is the Fourier transform of the ground state mass density profile of the nucleus,
normalized such that $F^2(q = 0) = 1$. The effective differential cross-section can
then be written as [5]:

$$\frac{d\sigma}{dE_r} = \frac{d\sigma_0}{dE_r} F^2(q) \quad \text{with} \quad q^2 = 2m_a E_r. \tag{2.4}$$

A form factor $F(q)$ is adopted as proposed by Klein and Nystrand [3] that is based
on the approximation of a Woods–Saxon nuclear density profile by a hard sphere of
radius R_A convoluted with a Yukawa potential of range a. The corresponding form
factor can then simply be written as the product of the Fourier transformation of
each individual density profile, that is:

$$F(q) = \frac{4\pi \rho_0 R_A^2}{Aq} j_1(q R_A) \frac{1}{1 + a^2 q^2} \tag{2.5}$$

$$= \frac{4\pi \rho_0}{Aq^3} (\sin(q R_A) - q R_A \cos(q R_A)) \frac{1}{1 + a^2 q^2} \tag{2.6}$$

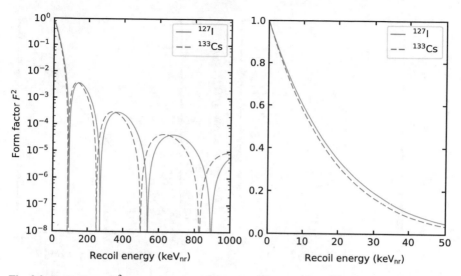

Fig. 2.2 Form factor F^2 versus recoil energy E_r for ^{127}I (blue) and ^{133}Cs (red). The left panel shows the form factor over a wide range of recoil energies. The right panel focuses on the recoil energies of interest to a CEνNS search at the SNS. A steep drop in F^2 for increasing recoil energies leads to a suppression of high-energy recoils in the resulting CEνNS spectrum

$$\text{with} \quad \rho_0 = \frac{3A}{4\pi R_A^3}, \quad a = 0.7\,\text{fm}, \quad R_A = A^{1/3} \cdot 1.2\,\text{fm} \tag{2.7}$$

where j_1 denotes the first-order spherical Bessel function of the first kind.

Figure 2.2 shows the form factor for cesium and iodine as calculated using Eq. (2.5) where the momentum transfer was converted into the experimentally more accessible recoil energy of the target nucleus. The left panel shows the behavior over a wide range of recoil energies whereas the right panel focuses on the ones expected at the SNS. There is a steep drop in F^2 for increasing recoil energies. According to Eq. (2.4), this directly leads to a heavy suppression of the differential cross-section for large recoil energies. The experimental consequence of this can be seen in Fig. 2.3. The left panel shows a comparison of the total CEνNS cross-section for ^{127}I, calculated with and without the inclusion of the form factor. For incoming neutrino energies of $E_\nu \gtrsim 30\,\text{MeV}$, the total cross-section starts to flatten out if F^2 is included. With increasing E_ν, the available phase space to integrate the differential cross-section increases, as $E_r^{\max} \propto E_\nu^2$. However, large recoil energies are heavily suppressed by the form factor, and as a result their contribution to the total cross-section is negligible. The right panel of Fig. 2.3 shows the average recoil energy induced in a hypothetical detector, which increases linearly with the incoming neutrino energy for $E_\nu \lesssim 30\,\text{MeV}$ but then again tapers off. The exact rate and shape of this tapering is solely determined by the form factor and as such the choice of $F(q)$ can significantly influence the overall recoil spectrum. It is therefore important to note that there exist several different form factor models which are

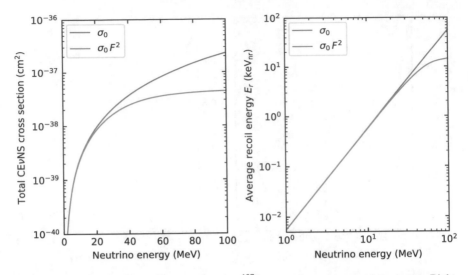

Fig. 2.3 Left: Total CEνNS cross-section for ^{127}I versus incoming neutrino energy. Right: Average recoil energy of ^{127}I versus incoming neutrino energy. Both panels include calculations with and without the inclusion of the form factor

based on different representations of the nucleon density. A more detailed discussion on different form factor models can be found in [4] and [6]. The latter reference also provides a more extensive and rigorous calculation of the CEνNS cross-section.

More detailed CEνNS cross-section calculations were carried out within the COHERENT collaboration including several different corrections to provide a precise SM prediction [7]. These include corrections such as the inclusion of axial vector currents and radiative corrections, which lead to an increase in the total cross-section on the order of 5–10% depending on the exact isotope. Different form factor models were tested ($\pm\mathcal{O}(5\%)$), as well as the impact of differences in the proton and neutron form factor (<1%). The inclusion of the strange quark radius was found to be negligible. The impact of strange quark contribution to the nuclear spin was found to be $\mathcal{O}(1\%)$ for low N, and is negligible otherwise. A small contribution of weak magnetism to the cross-section was also found to be negligible. The most important correction is the inclusion of different effective neutrino charge radii for different flavors [7, 8]. At low $q \approx 0$, this can be incorporated by replacing $\sin^2 \Theta_W$ with

$$\sin^2 \Theta_{\text{eff}} = \sin^2 \Theta_W + \frac{\alpha}{6\pi} \ln \left(\frac{m_i^2}{m_W^2} \right) \tag{2.8}$$

where m_i is the mass of the charged lepton associated with the neutrino ν_i. This immediately leads to different CEνNS cross-sections for different neutrino flavors, where the ratio between cross-section can approximately be written as:

$$\left. \frac{\sigma(\nu_i)}{\sigma(\nu_j)} \right|_{q^2=0} \approx 1 + \frac{\alpha}{3\pi \sin^2 \Theta_W} \ln\left(\frac{m_i^2}{m_j^2}\right). \tag{2.9}$$

The cross-section of muon neutrinos over electron neutrinos is therefore increased by a factor of $\sim$3.5%. This effect grants the possibility to distinguishing different neutrino flavors via neutral current interactions once precision measurements of the CEνNS cross-section are achieved.

References

1. A. Drukier, L. Stodolsky, Principles and applications of a neutral-current detector for neutrino physics and astronomy. Phys. Rev. D **30**(11), 2295 (1984)
2. J. Erler, M.J. Ramsey-Musolf, Weak mixing angle at low energies. Phys. Rev. D **72**, 073003 (2005). https://link.aps.org/doi/10.1103/PhysRevD.72.073003
3. S.R. Klein, J. Nystrand, Exclusive vector meson production in relativistic heavy ion collisions. Phys. Rev. C **60**, 014903 (1999). https://link.aps.org/doi/10.1103/PhysRevC.60.014903
4. J. Lewin, P. Smith, Review of mathematics, numerical factors, and corrections for dark matter experiments based on elastic nuclear recoil. Astropart. Phys. **6**(1), 87–112 (1996)
5. J. Engel, Nuclear form factors for the scattering of weakly interacting massive particles. Phys. Lett. B **264**(1), 114–119 (1991). http://www.sciencedirect.com/science/article/pii/037026939190712Y
6. D. Papoulias, T. Kosmas, Standard and nonstandard neutrino-nucleus reactions cross sections and event rates to neutrino detection experiments. Adv. High Energy Phys. **2015**, 763648 (2015)
7. P. Barbeau, D. Rimal, S. Klein et al., The coherent elastic ν-nucleus cross-section at the SNS. Technical report, Duke University (2015)
8. L. Sehgal, Differences in the coherent interactions of ν_e, ν_μ, and ν_τ. Phys. Lett. B **162**(4), 370–372 (1985). http://www.sciencedirect.com/science/article/pii/0370269385909426

Chapter 3
COHERENT at the Spallation Neutron Source

3.1 The Spallation Neutron Source

The Spallation Neutron Source (SNS), located at the Oak Ridge National Laboratory, provides a large group of interdisciplinary researchers with the most intense,
pulsed neutron beams in the world. A proton beam narrow in time with a timing
full width at half maximum (FWHM) of approximately $\Delta t_{\text{Beam}} \sim 380$ ns impinges
on a liquid mercury target with a repetition rate of 60 Hz. The SNS is theoretically
capable of delivering a total beam power on target of up to 1.4 MW. Combined
with a typical proton energy of $E_p \approx 1$ GeV, this amounts to a total proton rate of
$\Gamma_p \approx 10^{16}$ s^{-1}. Upon the impact on a target nucleus, protons only interact with an
individual nucleon instead of forming a compound, as their de Broglie wavelength
is only ~0.1 fm, and as such much smaller than the nucleus itself. Kinetic energy is
transmitted from a proton to the nucleon via elastic collisions after which an intra-
nuclear cascade ensues [1, 2]. During this nucleon-cascade, neutrons are spallated
from the target nucleus, and also pions $\pi^{\pm}$ are produced. The pions are stopped
within the target, where the π^- arc mostly recaptured by the mercury. The π^+,
in contrast, decay at rest into a positive muon and a muon neutrino. The muon
subsequently decays in-flight into a positron, an electron neutrino, and an anti-muon
neutrino.

After the initial nucleon-cascade, which lasts about 10^{-22} s, the nucleus is left in
a highly excited state. It then loses its remaining energy over $\sim 10^{-16}$ s mainly via
neutron evaporation. Over the course of the intra-nuclear cascade and the following
evaporation, an average of 20–30 neutrons per incoming proton are spallated from
the target [3]. Being produced in the intra-nuclear cascade, the timing of this neutron
emission is only associated directly with the beam. For the remainder of this thesis,
these beam-related neutrons are labeled as prompt. Additionally, a total of ~0.08
($\pm 10\%$) neutrinos per flavor per proton are produced during the full process [4]
where all neutrinos are emitted isotropically from the source.

© Springer Nature Switzerland AG 2018
B. Scholz, *First Observation of Coherent Elastic Neutrino-Nucleus Scattering*,
Springer Theses, https://doi.org/10.1007/978-3-319-99747-6_3

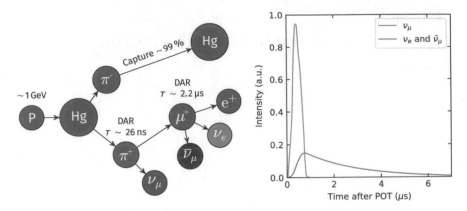

Fig. 3.1 **Left**: Neutrino production mechanism at the SNS. A proton beam impinges on a mercury target producing pions. Three different types of neutrinos are produced within the subsequent decay chain. Secondary neutrino emission cascades are not shown as they only contribute a negligible amount of contamination. **Right**: Timing profile of neutrino emission following protons-on-target (POT) as determined by the Florida group of the COHERENT collaboration using Geant4 simulations. A distinct difference in the arrival time is apparent between ν_μ, originating from pion decay, and the $\nu_e + \bar{\nu}_\mu$ emitted in the subsequent muon decay. The latter pair of neutrinos closely follows the distinct 2.2 μs muon decay profile

A schematic of the neutrino production mechanism is illustrated in the left panel of Fig. 3.1. The right panel shows the timing profile of the neutrino emission and further illustrates the timing profile of each neutrino species. The ν_μ are directly produced within the nucleon-cascade. As such, their time profile closely follows the proton beam profile. The ν_e and $\bar{\nu}_\mu$ emission follows a convolution of the beam profile with the ~2.2 μs muon decay. The neutrino emission can therefore be categorized as prompt (ν_μ) and delayed (ν_e, $\bar{\nu}_\mu$).

The pulsed neutrino emission profile is highly beneficial for background suppression. Potential CEνNS signals can only arise within a couple of microseconds directly following the protons-on-target (POT) trigger. As such, it is possible to reduce the steady-state contribution from environmental and cosmic-ray-induced radiation by approximately three to four orders of magnitude. However, the SNS also produces on the order of 10^{17} neutrons per second. Some of these can escape the shielding monolith surrounding the mercury target, leading to a potential beam-related background. This neutron background was assessed prior to the main detector deployment. It is described in Chap. 4 and found to be negligible.

The exact energy emission spectra for all neutrino flavors can be analytically calculated as all processes involved in the neutrino production are well understood. The $\pi^+ \rightarrow \mu^+ + \nu_\mu$ two-body decay-at-rest results in a monochromatic neutrino energy of

$$P(E_\nu|\nu_\mu) = \delta\left(E_\nu - \frac{m_\pi^2 - m_\mu^2}{2m_\pi}\right) = \delta\left(E_\nu - 29.79\,\text{MeV}\right) \tag{3.1}$$

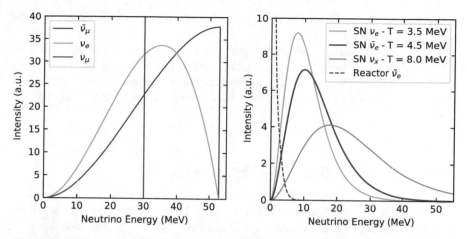

Fig. 3.2 Left: Neutrino spectra emitted from a stopped pion source, e.g., the SNS. **Right**: Typical neutrino spectra emitted during stellar core collapse (solid) [5] and emitted from a nuclear reactor (dashed) [6]. Here, ν_x denotes all other neutrino and anti-neutrino flavors. A large overlap between neutrino energies probed at the SNS and the emission from supernovae is apparent

The neutrino emission spectra for the $\mu^+ \to e^+ + \nu_e + \bar{\nu}_\mu$ decay can be written as [5]:

$$P(E_\nu | \nu_e) = \frac{12}{W^4} E_\nu^2 (W - E_\nu) \tag{3.2}$$

$$P(E_\nu | \bar{\nu}_\mu) = \frac{6}{W^4} E_\nu^2 \left(W - \frac{2}{3} E_\nu \right), \tag{3.3}$$

where $W = 52.83 \, \text{MeV}$. The energy spectrum for each flavor is shown in Fig. 3.2. The emitted neutrinos meet the coherence criterion ($E_\nu \leq 100 \, \text{MeV}$) introduced in Chap. 2. The right panel further shows the supernova neutrino emission spectra estimated using a Fermi–Dirac distribution with characteristic neutrino temperatures of $T(\nu_e) = 3.5 \, \text{MeV}$, $T(\bar{\nu}_e) = 4.5 \, \text{MeV}$, and $T(\nu_x) = 8 \, \text{MeV}$ [5, 7], that is:

$$f_{\text{FD}}(E_\nu, T) = \frac{E_\nu^2}{2T^3} \frac{1}{\exp^{E_\nu/T} + 1}. \tag{3.4}$$

The overlap in neutrino energies probed at the SNS and in supernovae are apparent. The CEνNS search at the SNS described in this thesis can therefore directly validate coherent scattering for neutrinos carrying energies similar to those produced in a stellar core collapse (Chap. 1). The reactor neutrino emission spectrum as provided in [6] is also shown. The emitted neutrino spectrum has a much lower energy than what is produced at the SNS. As such, there are some advantages to detect CEνNS-induced nuclear recoils at the SNS, even if the total recoil rate is significantly lower.

Geant4 [8] simulations were performed by the Florida group of the COHERENT collaboration regarding the neutrino yield, neutrino spectra, and timing profiles as provided by the SNS. A negligible energy contamination of <1% from decay-in-flight and muon capture was found above the endpoint of the Michel spectrum [4].

3.2   The COHERENT Experiment at the SNS

The work described in this thesis was done in the framework of the international COHERENT collaboration. COHERENT consists of approximately 80 researchers with strong backgrounds in rare event searches such as dark matter or $0\nu\beta\beta$-decay experiments. Yet, from the beginning it was clear that only a multi-target approach will be able to fully utilize the power of CEνNS to probe and constrain neutrino physics beyond the SM. The first observation of CEνNS described in this thesis is a remarkable achievement and already provided improved constraints on nonstandard interactions between neutrinos and quarks [4]. However, due to the low event statistics and the large uncertainties involved, many of the more interesting physics questions beyond a first observation cannot be addressed yet.

Using multiple target materials, the COHERENT experiment will further probe the N^2 dependance of the scattering cross-section as predicted by Eq. (2.3). Current operational CEνNS detectors include 14.57 kg of CsI[Na] as described in this thesis, a 28 $-$kg single-phase liquid argon (LAr) detector and 185 kg of NaI[Tl]. An additional CEνNS search using up to 20 kg of p-type point contact germanium detectors [9] is planned for deployment in the late 2017. Recent advancements enable a threshold of a few hundred keV$_{ee}$ in these detectors [10]. All of these setups are located in a basement corridor at the SNS, which is now dubbed "neutrino alley" (Fig. 3.3). This prime location provides at least 19 m of shielding against beam-associated backgrounds, such as prompt neutrons. It also offers an overburden of approximately 8 m.w.e. (meters of water equivalent) helping to further reduce cosmic-ray-induced backgrounds in the detectors.

In addition to the aforementioned detectors, COHERENT also deployed several different neutron monitors to verify the low prompt neutron background expected from the source. The neutron flux was measured at several positions within the neutrino alley (Fig. 3.3). In the near future, COHERENT will deploy another neutron monitor close to the CsI[Na] detector to further substantiate the current neutron flux measurements.

Another potential background was identified in the framework of the COHERENT collaboration. This background originates from neutrino-induced neutrons

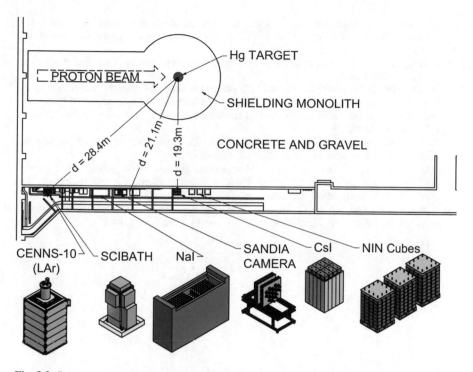

Fig. 3.3 Past, current, and future COHERENT detectors as positioned in a basement corridor at the SNS. The corridor is now dubbed 'neutrino alley' and provides at least 19 m of shielding against beam-related backgrounds. It further offers a total of 8 m.w.e. of overburden, helping to reduce cosmic-ray-induced backgrounds. The CENNS-10, NaI, and CsI detectors are designed to search for CEνNS signals. The Sandia Camera and Scibath on the other hand measured the prompt neutron rate at multiple positions along the alley. The NIN cubes are currently measuring the cross-section for neutrino-induced neutrons in Pb and Fe. The detector locations and their distance to the mercury target were determined using precision survey measurements performed by ORNL. These measurements are accurate to ±1 cm. Figure adapted from [4]

(NINs) in the lead shield surrounding the detectors. Accordingly, another detector array was deployed that is dedicated to measuring the cross-section of this process. A detailed description of these background measurements is presented in Chap. 4.

References

1. J. Cugnon, *Cascade Models and Particle Production: A Comparison* (Springer US, Boston, 1993), pp. 271–293. https://doi.org/10.1007/978-1-4615-2940-8_12
2. A. Krása, Spallation reaction physics. Czech Republic: Czech Technical University Lecture (2010). http://ojs.ujf.cas.cz/~krasa/ZNTT/SpallationReactions-text.pdf
3. C. Weiren, Spallation neutron source and other high intensity proton sources. Technical report, Fermi National Accelerator Lab (2003)

4. J.I. Collar, D. Akimov, J.B. Albert et al., Observation of coherent elastic neutrino-nucleus scattering. Science **357**(6356), 1123–1126 (2017). http://science.sciencemag.org/content/357/6356/1123
5. F. Avignone, Potential new neutrino physics at stopped-pion sources. Prog. Part. Nucl. Phys. **48**(1), 213–222 (2002)
6. T.A. Mueller, D. Lhuillier, M. Fallot et al., Improved predictions of reactor antineutrino spectra. Phys. Rev. C **83**, 054615 (2011). https://link.aps.org/doi/10.1103/PhysRevC.83.054615
7. M.T. Keil, Supernova neutrino spectra and applications to flavor oscillations (2003). Preprint. arXiv:astro-ph/0308228
8. S. Agostinelli, J. Allison, K.a. Amako et al., GEANT4 - a simulation toolkit. Nucl. Instrum. Methods Phys. Res. Sect. A **506**(3), 250–303 (2003)
9. P.S. Barbeau, J.I. Collar, O. Tench,   Large-mass ultralow noise germanium detectors: performance and applications in neutrino and astroparticle physics. J. Cosmol. Astropart. Phys. **2007**(09), 009 (2007). http://stacks.iop.org/1475-7516/2007/i=09/a=009
10. C.E. Aalseth, J. Colaresi, J.I. Collar et al., A low-noise germanium ionization spectrometer for low-background science. IEEE Trans. Nucl. Sci. **63**(6), 2782–2792 (2016)

Chapter 4
Background Studies

As discussed in the previous chapter, several different neutron detectors were deployed along the "neutrino alley" prior to the CsI[Na] experimentation. These measurements confirmed a negligible background level. This ensures a high signal-to-background level for the CEνNS search described in this thesis. These measurements are described below.

Two early measurements of the prompt neutron flux used the Scibath [1] and Sandia Camera [2] neutron detectors, which were positioned at several locations along the basement corridor as well as in the SNS instrument bay. A neutron flux of approximately 1.5×10^{-7} neutrons/cm^2 s(1–100 MeV) was measured in the vicinity of the location later used in the CsI[Na] deployment. However, these measurements carried a large uncertainty. To further constrain the prompt neutron flux and a potential NIN background, a neutron detector was positioned at the exact location later occupied by the CsI[Na] detector.

This neutron detector consisted of two 1.5 −liter liquid scintillator cells filled with EJ-301, which were housed in a shielding described in [3, 4]. The innermost layer of ultra-low background lead was removed to accommodate the detector cells. The shielding was further surrounded by an additional 3.5 in. of neutron moderator, i.e., aluminum tanks filled with water. The scintillator cells were read out by ET9390 PMTs.

Both scintillator and PMT used in this setup were selected primarily for their excellent neutron–gamma discrimination capability down to low energies [5, 6]. Using standard PSD techniques, neutron-like events were selected with arrival times ranging from 10 μs before the POT trigger to 10 μs after. Figure 4.1 illustrates the PSD used in this selection process. The energy range of these neutron-like events is limited from 30 to 300 keV$_{ee}$, where the lower bound is due to limitations in the n-γ discrimination capability and the upper bound is due to nonlinearities caused by PMT saturation. The distribution of event arrival times with respect to the POT trigger for neutron-like events in the EJ-301 is shown in Fig. 4.2. During the 171.7 days of experimentation, the SNS provided a total integrated beam

© Springer Nature Switzerland AG 2018
B. Scholz, *First Observation of Coherent Elastic Neutrino-Nucleus Scattering*,
Springer Theses, https://doi.org/10.1007/978-3-319-99747-6_4

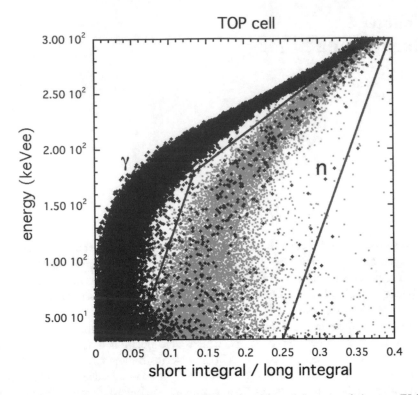

Fig. 4.1 Shown is the PSD calibration measurement performed for one of the two EJ-301 detectors. The scintillation decay times for nuclear and electronic recoils are different for EJ-301. The nature of events can be discriminated using a pulse shape discrimination parameter. The ratio between the scintillation light present in the beginning of an event to its total light output was chosen for this analysis. Shown above is the energy visible in the detector on the y-axis and the pulse shape discrimination parameter on the x-axis. The red band represents events taken in the presence of a γ-source and therefore represents electronic recoils. The blue band represents nuclear recoils in the presence of a neutron source. A clear separation between nuclear and electronic recoils is apparent for intermediate energies. However, at low energies the n-γ discrimination capability is limited as only a few photoelectrons are detected per event. For high energies, both bands merge because of nonlinearities caused by PMT saturation. Events within the blue contour are accepted as neutron-like. Figure courtesy of Juan Collar

power on the mercury target of 3.35 GWh. In the following sections, two beam-related backgrounds identified by the COHERENT collaboration are discussed: First, prompt neutrons originating from the mercury target (Sect. 4.1), and second, neutrino-induced neutrons (NINs) produced in the lead surrounding the CsI[Na] detector (Sect. 4.2).

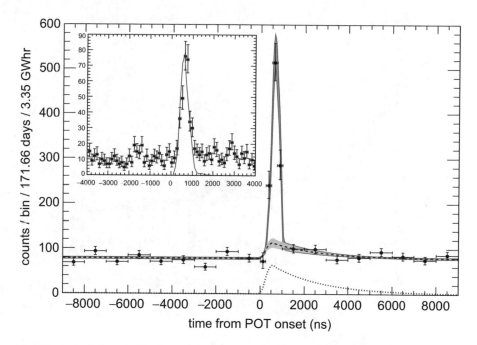

Fig. 4.2 Arrival time distribution of neutron-like events in the EJ-301 data set [7]. Red lines represent the 1σ confidence interval of the three-component model fit as described in the text. The dashed line represents the NIN component of the fit model. The dotted line shows the predicted NIN magnitude using rate predictions derived from [8, 9]. The inset shows the same data with a bin size of 100 ns. The red curve represents the normalized prompt neutron PDF as predicted from Geant4 simulations. Plot from [7]

4.1 Prompt Neutrons

A clear excess directly following the POT trigger is visible in Fig. 4.2. A prompt analysis window was defined from 200 to 1100 ns after the POT trigger. The event rate in this window was found to be approximately 0.7 events per SNS live-day.

The energy spectrum between 30 and 300 keV$_{ee}$ for all neutron-like events within this prompt window is shown in the upper panel of Fig. 4.3. Multiple comprehensive MCNPX-PoliMi ver. 2.0 simulations were conducted to determine the EJ-301 response to incoming neutron spectra of differing spectral hardness. In these simulations, the full detector and shielding geometry were unidirectionally exposed to neutrons coming from the SNS target. A power law, i.e., $P(E_n) \propto E_n[\text{MeV}]^{-\alpha}$, was chosen to model the spectral hardness of the incoming neutrons. This choice was informed by the previous measurements from the Scibath and Sandia Camera neutron detectors. The neutron energies simulated ranged from 1 to 100 MeV. Neutrons with energies below 1 MeV lose too much energy within the neutron moderators surrounding the EJ-301 cells and do not contribute significantly to the

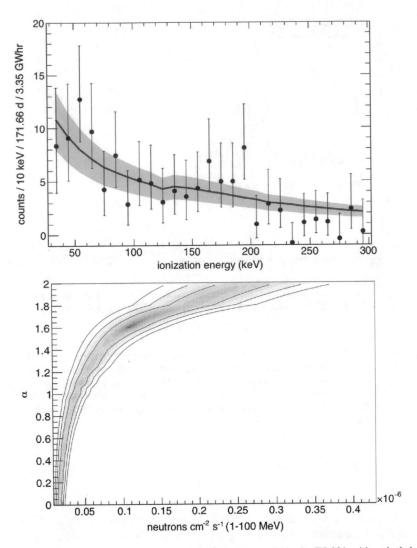

Fig. 4.3 Top: Spectrum of neutron-like energy depositions within the EJ-301 with arrival times consistent with prompt neutrons, i.e., 200–1100 ns after the POT trigger. Contributions from environmental neutrons were removed from the spectrum using neutron-like events in anti-coincidence with the POT trigger, i.e., arrival times of less than zero in Fig. 4.2. The blue line represents the simulated EJ-301 response to the best-fit neutron spectral model tested. The blue-shaded band shows the energy range corresponding to all parameter combinations within the 1σ band of the bottom panel. **Bottom**: Goodness-of-fit for all neutron models tested. Darker areas represent better fits. The red lines show the $1-3\sigma$ levels of the fit. Plot from [7]

30–300 keV$_{ee}$ energy range. The flux of neutrons with energies above 100 MeV was found to be negligible by the Scibath and Sandia Camera neutron detectors.

The MCNPX-PoliMi ver. 2.0 output was post-processed to incorporate the response of EJ-301 to nuclear recoils. The recoil energies were converted into an electron equivalent energy using the known quenching factors for hydrogen and carbon [10–13]. The calculation of the total scintillation light produced in an event also included statistical fluctuations in the light generation.

The simulated energy spectrum were fitted to the experimental data by varying the neutron flux ϕ_n and spectral hardness α. The goodness-of-fit between the simulated detector response and the experimental energy spectrum is shown in Fig. 4.3. The spectral hardness α is shown on the y-axis, whereas the corresponding neutron flux ϕ_n is shown on the x-axis. A darker area represents a better agreement between the simulated and measured energy spectra. Red lines represent the corresponding $1 - 3\sigma$ confidence intervals. The best fit is given by:

$$\alpha = 1.6 \quad \text{and} \quad \phi_n = 1.09 \times 10^{-7} \frac{n}{\text{cm}^2\,\text{s}}. \tag{4.1}$$

The energy spectrum calculated using this parameter set is shown as a blue line in the upper panel of Fig. 4.3. The blue band consists of all parameter choices covered within the 1σ confidence interval. The calculated neutron flux is compatible with previous flux measurements using the Scibath and Sandia Camera neutron detectors, which measured a prompt neutron flux of 1.5×10^{-7} neutrons/cm^2 s(1–100 MeV) close to the position occupied by the EJ-301 detector.

A second MCNPX-PoliMi ver. 2.0 simulation was used to estimate the beam-related background rate caused by prompt neutrons in the CsI[Na] detector during this CEνNS search. A comprehensive model of the CsI[Na] setup was uniformly bathed in neutrons coming from the SNS target. The neutron spectral hardness and flux were fixed to the values given in Eq. (4.1). During the post-processing of the simulation output, the proper CsI[Na] response to nuclear recoils was used, i.e., light-yield and quenching factor as calculated later in this thesis (Chaps. 6 and 8). Poisson fluctuations were added to the number of photoelectron (PE) generated during an event. An energy spectrum was calculated from the simulated event energies and the signal acceptance model corresponding to the optimized choice of cut parameters used in the CEνNS search (Chap. 9) was applied to the spectrum. The uncertainties of light yield (Chap. 6), quenching factor (Chap. 8), and signal acceptance (Chap. 9) were properly propagated through this analysis exercise. The resulting background rate from prompt neutrons was found to be

$$\Gamma_{\text{prompt}} = 0.92 \pm 0.23 \frac{\text{events}}{\text{GWh}}. \tag{4.2}$$

The arrival time of the background events caused by prompt neutrons is highly concentrated at 200–1100 ns after the POT trigger (Fig. 4.2). In Chap. 9, it is shown that this event rate is approximately 25 times smaller than the expected CEνNS signal rate.

4.2   Neutrino-Induced Neutrons (NINs)

As discussed earlier, neutrons produced by neutrinos interacting with lead [8, 9] constitute another relevant background. These NINs are produced by the charged-current reaction:

$$\nu_e \quad + \quad {}^{208}\text{Pb} \quad \rightarrow \quad {}^{208}\text{Bi}^* \quad + \quad e^-$$

$$\downarrow$$

$$^{208-y}\text{Bi} \quad + \quad x \times \gamma \quad + \quad y \times n$$

and the neutral-current interaction:

$$\nu_x \quad + \quad {}^{208}\text{Pb} \quad \rightarrow \quad {}^{208}\text{Pb}^* \quad + \quad \nu_x'$$

$$\downarrow$$

$$^{208-y}\text{Pb} \quad + \quad x \times \gamma \quad + \quad y \times n.$$

Here, x and y denote the γ and neutron multiplicity, respectively. The NIN background is predominantly produced by the delayed ν_e, as the charged-current interaction has the largest cross-section. An unbinned fit [14] to the EJ-301 arrival time data (Fig. 4.2) was used to constrain this NIN background. The model used in this fit incorporated the following three neutron components: a random arrival time from environmental neutrons, prompt neutrons, and a NIN excess following the ν_e time profile. The number of NINs found in this fit was converted into a NIN production rate using an MCNPX-PoliMi ver. 2.0 simulation. A homogeneous and isotropic NIN emission within the lead shield surrounding the EJ-301 was simulated. An energy spectrum corresponding to the highest stellar-collapse neutrino energies ($T = 8\,\text{MeV}$) described in [15] is adopted for the NIN emission. The energy spectrum of these stellar-collapse neutrinos is slightly softer than the energy spectrum of ν_e at the SNS (Fig. 3.2). However, the change in NIN spectral hardness is negligible for different neutrino energies [15]. The energy spectrum measured by the EJ-301 was calculated using an identical approach to the one described in Sect. 4.1. The total NIN production rate was found to be

$$\Gamma^\star_{\text{nin}} = 0.97 \pm 0.33 \frac{\text{neutrons}}{\text{GWh kg of Pb}}. \tag{4.3}$$

An additional simulation was performed using homogeneous and isotropic NIN emission in the lead surrounding the CsI[Na] used in the CEνNS search and analyzed as described in Sect. 4.1. The uncertainties of light yield (Chap. 6), quenching factor (Chap. 8), and signal acceptance (Chap. 9) were properly propagated through this exercise. Scaling the simulated NIN event rate with the expected NIN production rate resulted in a final background rate of

$$\Gamma_{nin} = 0.54 \pm 0.18 \, \frac{\text{events}}{\text{GWh}}. \tag{4.4}$$

In Chap. 9, it is shown that this event rate is approximately 43 times smaller than the expected CEνNS signal rate.

These calculations indicate that both prompt neutrons and NINs only contribute a negligible background to the CEνNS search. The validity of the neutron transport simulations is further substantiated in Sect. 9.3.1.

References

1. S.J. Brice, R.L. Cooper, F. DeJongh et al., A method for measuring coherent elastic neutrino-nucleus scattering at a far off-axis high-energy neutrino beam target. Phys. Rev. D **89**, 072004 (2014). http://link.aps.org/doi/10.1103/PhysRevD.89.072004
2. N. Mascarenhas, J. Brennan, K. Krenz et al., Results with the neutron scatter camera. IEEE Trans. Nucl. Sci. **56**(3), 1269–1273 (2009)
3. J. Collar, N. Fields, M. Hai et al., Coherent neutrino-nucleus scattering detection with a CsI[Na] scintillator at the SNS spallation source. Nucl. Instrum. Methods Phys. Res. Sect. A **773**, 56–65 (2015). http://www.sciencedirect.com/science/article/pii/S0168900214013254
4. N.E. Fields, CosI: development of a low threshold detector for the observation of coherent elastic neutrino-nucleus scattering. Ph.D. thesis, 2014. https://search.proquest.com/docview/1647745354?accountid=14657
5. X. Luo, V. Modamio, J. Nyberg et al., Test of digital neutron–gamma discrimination with four different photomultiplier tubes for the NEutron Detector Array (NEDA). Nucl. Instrum. Methods Phys. Res. Sect. A **767**, 83–91 (2014)
6. E. Ronchi, P.-A. Söderström, J. Nyberg et al., An artificial neural network based neutron–gamma discrimination and pile-up rejection framework for the BC-501 liquid scintillation detector. Nucl. Instrum. Methods Phys. Res. Sect. A **610**(2), 534–539 (2009)
7. J.I. Collar, D. Akimov, J.B. Albert et al., Observation of coherent elastic neutrino-nucleus scattering. Science **357**(6356), 1123–1126 (2017). http://science.sciencemag.org/content/357/6356/1123
8. R. Lazauskas, C. Volpe, Low-energy neutrino scattering measurements at future spallation source facilities. J. Phys. G: Nucl. Part. Phys. **37**(12), 125101 (2010). http://stacks.iop.org/0954-3899/37/i=12/a=125101
9. R. Lazauskas, C. Volpe, Corrigendum: Low-energy neutrino scattering measurements at future spallation source facilities (2010 J. Phys. G: Nucl. Part. Phys. 37 125101). J. Phys. G: Nucl. Part. Phys. **42**(5), 059501 (2015). http://stacks.iop.org/0954-3899/42/i=5/a=059501
10. D. Ficenec, S. Ahlen, A. Marin et al., Observation of electronic excitation by extremely slow protons with applications to the detection of supermassive charged particles. Phys. Rev. D **36**(1), 311 (1987)
11. V. Verbinski, W. Burrus, T. Love et al., Calibration of an organic scintillator for neutron spectrometry. Nucl. Instrum. Methods **65**(1), 8–25 (1968). http://www.sciencedirect.com/science/article/pii/0029554X68900037
12. S. Yoshida, T. Ebihara, T. Yano et al., Light output response of KamLAND liquid scintillator for protons and 12 C nuclei. Nucl. Instrum. Methods Phys. Res. Sect. A **622**(3), 574–582 (2010)
13. J. Hong, W. Craig, P. Graham et al., The scintillation efficiency of carbon and hydrogen recoils in an organic liquid scintillator for dark matter searches. Astropart. Phys. **16**(3), 333–338 (2002)

14. W. Verkerke, D. Kirkby et al., The RooFit toolkit for data modeling, in *Statistical Problems in Particle Physics, Astrophysics and Cosmology*, ed. by L. Lyons, M.K. Ünel (Imperial College Press, London, 2006), pp. 186–190
15. E. Kolbe, K. Langanke, Role of ν-induced reactions on lead and iron in neutrino detectors. Phys. Rev. C **63**, 025802 (2001). https://link.aps.org/doi/10.1103/PhysRevC.63.025802

Chapter 5
The CsI[Na] CEνNS Search Detector at the SNS

5.1 Overview and Wiring of the Detector Setup

A schematic of the CsI[Na] detector setup deployed at the SNS is shown in Fig. 5.1. The assembly is located 19.3 m from the SNS mercury target in a basement corridor (Fig. 3.3). The central detector is a 34 −cm-long sodium-doped cesium iodine scintillator read out by a super-bialkali R877-100 PMT by Hamamatsu. The detector is surrounded by 3 in. of high-density polyethylene (HDPE). The purpose of this innermost layer of PMT was to reduce the background caused by NIN production in the surrounding lead (Chap. 4). This layer of HDPE is followed by 2 in. of low-background lead with an approximate ^{210}Pb contamination of ~10 Bq kg^{-1}. The outer γ-shield is made of an additional 4 in. of contemporary lead (~100 Bq kg^{-1} of ^{210}Pb). This results in a minimum of 6 in. of lead surrounding the CsI[Na] in any direction with a total mass of approximately 3.5 tons. The lead shield is encased by a 2-in.-thick high-efficiency muon veto on all vertical sides and the top. The setup rests upon a platform made of an additional 6 in. of HDPE selected for its low-background properties. A heavy-duty frame made of aluminum Bosch extrusions is placed around the setup providing a secure anchor point for the muon veto panels. It provided a firm surface against which an additional layer of aluminum tanks was pushed, without disturbing the inner shielding configuration. These tanks were filled with tap water, providing at least an additional 3.5 in. of neutron moderator. Figure 5.2 shows the setup at three different stages during the installation process in June 2015.

A schematic of the data acquisition system is shown in Fig. 5.3. The high voltage of +1350 V for the central CsI[Na] detector is provided by an Ortec 556 power supply. The PMT output signal is fed into a Phillips Scientific (PS) 777 variable gain amplifier. The DC level of the signal is shifted by approximately +90 mV and amplified using the PS 777. The DC shift was necessary to make use of the full data acquisition range. The additional amplification increased the SPE amplitude without increasing the baseline noise level, avoiding the need to operate the PMT

© Springer Nature Switzerland AG 2018
B. Scholz, *First Observation of Coherent Elastic Neutrino-Nucleus Scattering*,
Springer Theses, https://doi.org/10.1007/978-3-319-99747-6_5

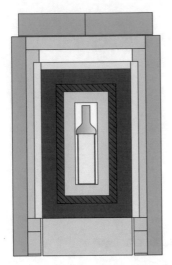
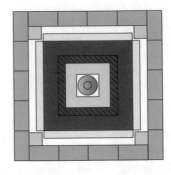

Fig. 5.1 Schematic of the CsI[Na] shielding at the SNS. The shielding components from the inside out are as follows: (1) light gray: 3 in. of low-background HDPE. (2) Hatched dark gray: 2 in. of low-background lead. (3) Dark gray: 4 in. of contemporary lead. (4) Yellow: 2-in.-thick muon veto panels. (5) Green: Aluminum Bosch-Rexroth extrusions. (6) Blue: Aluminum tanks filled with water, on the sides, WaterBricks, on the top

Fig. 5.2 Pictures of the CEνNS search detector deployment in June 2015 at the SNS. The leftmost picture highlights the innermost layers of HDPE used to reduce the background originating from NIN production in the surrounding lead (Chap. 4). The middle picture shows the setup once the aluminum frame was assembled. The right picture shows the full installation, including the NIM rack. Images courtesy of Juan Collar

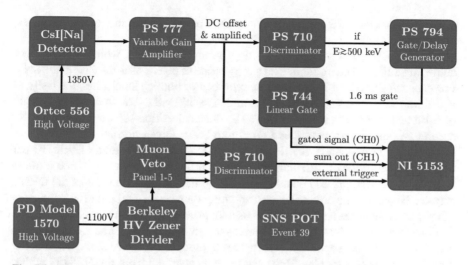

Fig. 5.3 Schematic of the data acquisition system used during the CEνNS search

at an excessive high voltage. The PS 777 output is split, one output is fed into a PS 710 discriminator and the other one into a PS 744 linear gate. The PS 710 provides a logical signal to a PS 794 gate/delay generator whenever the CsI[Na] signal contains an energy deposition of $\gtrsim$500 keV$_{ee}$. The PS 794 provides a ~1.6-μs-long logical signal which closes the PS 744 linear gate. The output of the PS 744 is fed into channel 0 of the National Instruments (NI) 5153 fast digitizer.

The linear-gate logic, consisting of the PS 710, 794 and 744, was necessary as any ungated high-energy deposition in the CsI[Na], caused, for example, by a traversing muon, would reset the data acquisition card, rendering the data acquisition system unresponsive for approximately 3 s. The linear-gate logic prevents this reset and enables a continuous data acquisition. A gate length of 1.6 ms was chosen in order to cut most of the afterglow of any high-energy event while only introducing a minimal dead time of only $\mathcal{O}(1\%)$.

The high voltage for all muon veto panels is provided by a Power Designs Model 1570, which is set to -1100 V. The 1570 output is fed into a Berkeley HV Zener Divider, which in return provides nine distinct high-voltage output levels, one for each PMT used within the muon veto. By operating each PMT at an individual high voltage, it was possible to match the gain of all PMTs (Fig. 5.8). The output of the five muon veto panels, i.e., four side panels and one top panel, is fed into the PS 710 discriminator. The summed, discriminated output of all panels is fed into channel 1 of the NI 5153. The external trigger for the data acquisition system is the *event 39* signal provided by the SNS. *Event 39* is generated when there is a potential extraction of protons from the storage ring, even if there are no protons stored in the ring, which approximately happens once in every 600 triggers. As such, *event 39* provides a stable 60 Hz trigger signal with an amplitude of ~2.5 V for all times no matter whether the SNS actually provided POT or not. During the rest of this thesis, this triggering signal is referred to as POT trigger.

A LabVIEW-based data acquisition program was written, which is capable of achieving a 100% data throughput at a trigger rate of 60 Hz, as is the case at the SNS. For each trigger 70-μs-long traces are recorded of both channels, i.e., the CsI[Na] and muon veto. The waveforms are sampled at 500 MS s^{-1} with a digitizer depth of 8-bit and saved as raw binary files. The data acquisition software automatically compresses the binary data files into zip-archives. Given normal operations, this amounts to approximately 60 MB per minute or a manageable total of $\sim$90 GB per day. An in-depth look at the data structure is provided in Sect. 5.4. Once a run is completed, the compressed data is pushed onto the HCDATA cluster of the ORNL Physics Division. The cluster keeps a spinning-disk copy of all data for fast access during the actual data analysis as well as a long-term archive on tape.

Once the CsI[Na] detector was deployed, an unexpected, minor complication was found in the data acquisition software. In order to minimize storage space, the data is saved as I8 variables. However, the data acquired from the NI 5153 within LabVIEW is actually provided as double precision floating point numbers in mV. These floating point values are converted back to their 8 $-$bit depth by applying

$$V_{i8} = 2^8 \times \frac{V_{dbl}}{V_{Range}}, \tag{5.1}$$

where V_{dbl} is the digitizer value in mV, V_{Range} the current digitizer range, and V_{i8} the digitizer value in ADC counts. Once a first batch of data had been analyzed, it became apparent that V_{Range} slightly differs from the setting provided, i.e., a digitizer range set to 0.2 V_{pp} actually shows a slightly larger range. This issue was confirmed by an NI technician. The projection of V_{dbl} onto V_{i8} therefore resulted in some I8 values that never appear as no initial V_{dbl} is mapped onto them. A correction needs to be applied during the analysis that maps the I8 values onto a continuous signal amplitude. This does not have any ill effect on the results presented here. Once this issue had been identified, the impact of this on the data analysis was investigated. Some I8 values are simply omitted, and as such the effective digitizer range is reduced by approximately 20 ADC counts. But, as all further analysis presented in this thesis was based on ADC counts and not an absolute value in mV, the analysis remained unaffected by this issue.

The exact transformation depends on V_{Range}. For the two digitizer ranges used in this thesis, the transformations that map the recorded V_{i8} data onto a continuous set of signal amplitude values V are given by:

$$V(V_{Range} = 200\,\text{mV}_{pp}) = \text{int}\,(V_{i8}) - \text{int}\left(\text{floor}\left(\frac{\text{double}\,(V_{i8}) + 5.0}{11.0}\right)\right) \tag{5.2}$$

$$V(V_{\text{Range}} = 500\,\text{mV}_{\text{pp}}) = \text{int}\,(V_{i8}) + \text{sign}\,(V_{i8})\,\text{int}\left(\text{floor}\left(\frac{4.0 - \text{double}\,(V_{i8})}{11.0}\right)\right)$$

$$(5.3)$$

Here, double() and int() represent data type casts into double precision floating point numbers and integers, respectively. The floor() function returns the largest integer less than or equal to its argument.

5.2 The Central CsI[Na] Detector

The inorganic scintillator CsI[Na] was chosen as the target material due to its many beneficial properties simplifying a CEνNS detection and making it possible in the first place [1]. The crystal was grown by AMCRYS-H in Ukraine, and is approximately 34 −cm long with a total mass of 14.57 kg. It is wrapped in an expanded PTFE membrane reflector and encapsulated in an electroformed OFHC copper can, which was grown at Pacific Northwest National Laboratory (PNNL). Figure 5.4 shows the encapsulated detector during its installation at ORNL. Both ^{133}Cs and ^{127}I nuclei have a large number of neutrons, 78 and 74, respectively, which leads to a large coherent enhancement of the scattering cross-section, as shown in Eq. (2.3). The large size of the nucleus also allows for a larger momentum transfer between the incoming neutrino and the target nucleus before a loss of coherence sets in as governed by the form factor (Eq. (2.5)). Another benefit of this detector material is the very similar nuclear mass of cesium and iodine. As such, the CEνNS-induced recoil spectra are nearly indistinguishable from one another, leading to a much simpler data analysis. Yet, given the modest neutrino energies and the relatively large mass of the recoiling nucleus, it is still necessary to achieve a low enough energy threshold to actually benefit from this enhancement.

In order to achieve such a low-energy threshold, an R877-100 5-in. flat-faced PMT from Hamamatsu was chosen to read out the scintillation light. This PMT uses a super-bialkali (SBA) photocathode which provides an increased quantum efficiency (QE) of ∼34% over the ∼24% QE of common bialkali photocathodes [2]. The SBA photocathode is further matched to the scintillation light of CsI[Na] as is shown in the left panel of Fig. 5.5. The high QE, paired with a light yield of CsI[Na] of ∼45 photons per keV$_{\text{ee}}$ [3] made it possible to achieve an energy threshold of approximately 4.5 keV$_{\text{nr}}$. CsI[Na] has the additional advantage of short scintillation decay times (Fig. 5.6), as well as a manageable afterglow (phosphorescence), especially when compared to thallium-doped CsI (Fig. 5.7). As such, the detector can be operated at a modest overburden without contaminating a significant amount of SNS proton-on-target triggers with SPEs from previous cosmic-ray-induced events.

The right panel of Fig. 5.5 highlights yet another advantage of CsI[Na], namely the excellent light yield stability around room temperature. A change of ±10 °C only results in a negligible change of its light yield on the order of less than 1%.

Fig. 5.4 Image of the central CsI[Na] detector during installation. From left to right: Author of this thesis, Grayson Rich (University of North Carolina at Chapel Hill). Image courtesy of Juan Collar

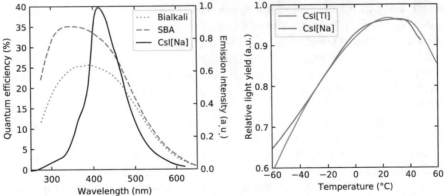

Fig. 5.5 Left: Quantum efficiency of bialkali and super-bialkali (SBA) photocathodes are shown in red and blue respectively [2]. The emission profile of sodium-doped CsI is shown in black [3]. The match between the quantum efficiency of the photocathodes and the scintillator emission profile is evident. **Right:** Temperature stability of CsI[Na] and CsI(Tl). The minimal change in light yield between 10 and 30 °C is readily visible. Adopted from [4]

This has been highly beneficial as over the course of the last 2 years several other detectors were deployed within the *neutrino alley*. As a result, the heat load was always changing which leads to fluctuations in the environmental temperature on the order of $\pm 2\,°C$. Yet, the excellent temperature stability guaranteed a constant light emission throughout the full data run. It was also shown that CsI[Na] crystals show very low levels of internal radioactivity, i.e., ^{238}U, ^{232}Th, ^{40}K, ^{126}I, and $^{134,\,137}$Cs,

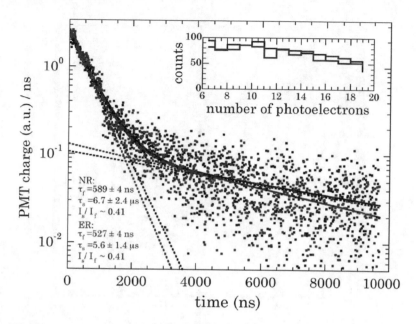

Fig. 5.6 Electronic and nuclear decay times of low-energy events in CsI[Na] [1]. The decay profile was decomposited into a fast and slow component [5]. CsI[Na] shows almost the same decay profile for both recoil types, which greatly simplifies the detector calibration (Chap. 7). The $60-ns$ difference in the fast decay component might grant us the ability to statistically differentiate between nuclear and electronic recoils in the future. Figure courtesy of Juan Collar

when grown from appropriate salts [6]. In her PhD thesis [7], Nicole Fields screened a boule slice of CsI[Na] from AMCRYS-H using low-background γ spectroscopy. She measured an activity per isotope on the order of $\mathcal{O}(10\,mBq\,kg^{-1})$. An in-depth analysis of the salts used by AMCRYS-H to grow the CsI[Na] used in this thesis, and most other materials used during the construction of this detector setup are provided in Chapter 4 of [7].

5.3 The Muon Veto

The muon veto panel consists of four side panels and one top panel, each made out of EJ-200B. The dimension for each side panel is $42 \times 24 \times 2$ in., the top panel is $24 \times 24 \times 2$ in. Each side panel is read out by two DC-chained, gain-matched ET 9102SB PMTs with a diameter of 1.5 in. The top panel is read out by a single 5-in. flat faced PMT.

Before assembly, the gain curve of each ET 9102SB was mapped by attaching each PMT to the same crystal and irradiating the setup with a ^{57}Co source, which provides γs with an energy of 122 keV. An Amptek MCA 8000A (Pocket MCA)

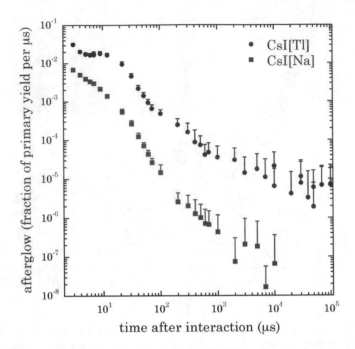

Fig. 5.7 Afterglow (phosphorescence) of CsI[Na] and CsI[Tl] as a fraction of primary scintillation yield [1]. The afterglow experienced by CsI[Na] is not as excessive as the afterglow in CsI[Tl]. Given the modest overburden of 8 m.w.e. available at the CEνNS search, this was crucial as otherwise most triggers would contain a significant number of SPEs from phosphorescence following high-energy, cosmic-ray-induced events. Figure courtesy of Juan Collar

was used to record the energy spectrum for each PMT operated at several different high voltages. The ^{57}Co peak position was recorded for each of these spectra. The resulting gain curves are shown in Fig. 5.8. Once each muon veto panel was wrapped with reflecting and light-blocking material, the separation between the environmental γ background and the high-energetic muon interactions was tested for different gain levels. The final operational high voltage for each PMT is listed in Table 5.1.

The PS 710 discriminator level for each panel is listed in Table 5.1. The levels for all side panels were set to achieve a maximum muon veto rejection while minimizing the triggering on the environmental γ background. All side panels showed a similar triggering rate of 20 Hz. The top panel showed a triggering rate of approximately 50 Hz. All rates are in agreement with rate expectations based on the muon flux at sea level given the veto surface area. The individual panel locations with respect to the central CsI[Na] detector are shown in Fig. 5.9, which also includes the Mercury Off-Gas Treatment System (MOTS) exhaust pipe [8] running along the ceiling of the corridor. This pipe is a major source of γ-rays, mainly carrying an energy of 511 keV. As such, the overall triggering rate of all panels was found to be slightly elevated whenever the SNS was operational. The

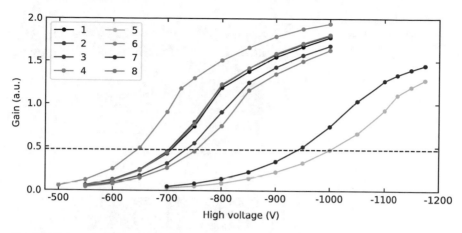

Fig. 5.8 Gain curves for each PMT used in the CsI[Na] muon veto. The gain was determined using a ^{57}Co source and a Pocket MCA. The dashed black line represents the gain to which all PMTs were matched. The high voltage chosen for each PMT is given in Table 5.1

Table 5.1 High voltages for all PMTs used in the CsI[Na] muon veto

PMT No.	Panel	High voltage (V)	Disc. level (mV)	Serial No.
1	1	−700	14.0	107027
2	1	−700		107010
3	2	−720	14.5	106958
4	2	−740		64454
5	3	−1000	13.7	64514
6	3	−660		64425
7	4	−940	13.2	64463
8	4	−700		64467
9	5	−1100	25.2	N/A

Also shown is the panel number each PMT is attached to, as well as the discriminator level used for each panel

increase in rate depends on the exact panel positioning. Panel five experienced the largest increase with an overall triggering rate of ∼100 Hz, whereas the side panels only experienced an increase in triggering rate on the order of 2–8 Hz, depending on their positioning.

5.4 Data Structure

The folder structure created by the LabVIEW DAQ system is shown in Fig. 5.10. The main folder is used to distinguish between different setups, e.g., calibrations or CEνNS search runs. The run folders contain the year, month, day, 24-h, minute,

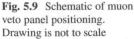

Fig. 5.9 Schematic of muon
veto panel positioning.
Drawing is not to scale

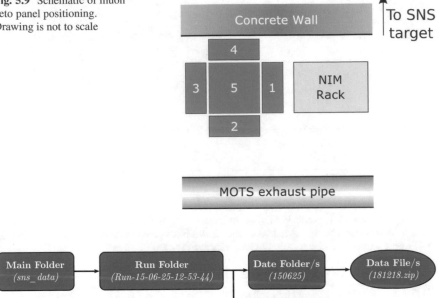

Fig. 5.10 The main folder distinguishes between different detector setups. The run folder contains
the time when the run was actually started. A new date folder is created every midnight. A new
data file is created every minute per default

and second in which this specific run was started. The date folders consist of
six digits, representing year, month, and day. During data acquisition, a new date
folder is created every midnight. A new data file is being created every minute
per default.

During development, both hardware and software were optimized to allow for
a simultaneous data acquisition and data compression, without introducing any
throughput limitations. The compression of a 1 min binary data file with an initial
size of 240 MB into a zip-archive takes approximately 45 s. As such, no data pile
up is being introduced by compressing the data. This is highly beneficial for data
storage as the compressed data only requires approximately a quarter of the disk
space of the initial binary files. It also reduces the time needed to back up the data
on the HCDATA cluster. Given a data transfer rate of 5–10 MB s^{-1}, raw binary files
could be transferred without introducing any data pile up. However, these transfer
rates fall below 5 MB s^{-1} for prolonged times on a regular basis. As such, new data
would be created faster than it can be moved onto the cluster. As such, the data

Table 5.2 Description for all data values available in the settings file of each data run

Index	Description	Index	Description
0	CH0 Coupling [AC = 0, DC = 1]	13	Trigger Source [CH0, CH1, Ext = 3]
1	CH0 Probe Attenuation	14	Trigger Level [V]
2	CH0 Vertical Range [V]	15	Hysteresis [V]
3	CH0 Vertical Offset [V]	16	Holdoff [s]
4	CH1 Coupling [AC = 0, DC = 1]	17	Trigger Delay [s]
5	CH1 Probe Attenuation	18	Trigger Type [Edge = 0,Hyst = 1]
6	CH1 Vertical Range [V]	19	-reserved-
7	CH1 Vertical Offset [V]	20	-reserved-
8	Minimum Sample Rate [S/s]	21	-reserved-
9	Minimum Record length [S]	22	Start time stamp[a]
10	Trigger Reference Position [%]	23	End time stamp[a]
11	Trigger Coupling [AC = 0, DC = 1]	24	-reserved-
12	Triggering Slope [neg = 0, pos = 1]	25	-reserved-

[a] LabView not Unix, i.e., whole seconds after the Epoch 01/01/1904 00:00:00.00 UTC

acquisition would have to be stopped once the local hard drive is at full capacity. Being able to compress the data in parallel to the data acquisition fully eliminates this issue.

The settings file contains all DAQ settings used during the acquisition. A full description of each value saved is presented in Table 5.2. Some parameters are saved in the settings file even though they have no effect on the NI 5153 digitizer, e.g., the vertical offset of either channel. These parameters are technically available in the sub-VIs provided by NI to properly initialize the hardware, yet internally the NI 5153 digitizer is not capable of adjusting these values. Fields marked as -reserved- do not contain any meaningful parameters. They might contain nonzero entries as they were used during the development of the DAQ software. Yet, at this stage the values provided do no longer carry any meaningful information and can be disregarded.

As already stated, the acquired data is saved as binary files containing an array of I8 variables. Figure 5.11 shows the internal structure of an example binary data file. Using LabVIEW for the data analysis would provide built-in functions to properly read the chunks of data and decimate the corresponding arrays. However, due to the sheer size of the data set acquired the full analysis had to be performed on the HCDATA cluster, fully utilizing its potential of parallel data analysis. The analysis code used in this thesis is available at [9].

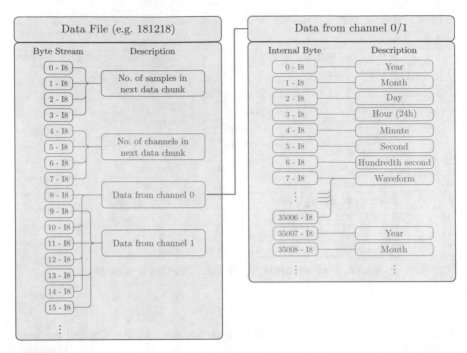

Fig. 5.11 Shown is the internal structure of the binary files saved by the LabView DAQ. Each binary file (left) consists of multiple, successive chunks of data. The first four bytes (I8, 0-3, red in the figure) of each chunk have to be merged into a single U32 which is the number of samples N_S in each channel that belong to the current chunk. For the measurements presented in this thesis (^{241}Am, ^{133}Ba, and SNS), N_S has to be a multiple of 35,007. The second quartet of bytes (I8, 4-7, green in the figure) also has to be merged into a single U32 which is the number of channels N_C in the current chunk. N_C will always be 2 for this DAQ. Next, a total of $N_S \times N_C$ bytes follow which have to be decimated into two arrays. One array contains all data belonging to CH0 within this chunk and the other one contains all data belonging to CH1. One of these arrays is shown on the right. Each array consists of $N = N_S/35,007$ subsets of 35,007 bytes. The first 7 bytes of each subset contain all the timing information needed for an event. The following 35,000 bytes represent the waveform. The next LabView *header* is encountered at byte index $i = N_S \times N_C + 8$ of the binary data file, where the procedure described above has to be repeated in order to extract the individual waveforms from the chunk

References

1. J. Collar, N. Fields, M. Hai et al., Coherent neutrino-nucleus scattering detection with a CsI[Na] scintillator at the SNS spallation source. Nucl. Instrum. Methods Phys. Res. Sect. A **773**, 56–65 (2015). http://www.sciencedirect.com/science/article/pii/S0168900214013254
2. H.P.K.K., Photomultiplier tubes and assemblies. For scintillation counting and high energy physics (2000). http://www.hamamatsu.com/resources/pdf/etd/High_energy_PMT_TPMZ0003E.pdf
3. J.T.M. de Haas, P. Dorenbos, Advances in yield calibration of scintillators. IEEE Trans. Nucl. Sci. **55**(3), 1086–1092 (2008)

4. CsI(Tl), CsI(Na), Cesium iodide scintillation material (2016). http://www.crystals.saint-gobain. com/sites/imdf.crystals.com/files/documents/csitl-and-na-material-data-sheet_69771.pdf
5. S. Keszthelyi-Lándori, G. Hrehuss, Scintillation response function and decay time of CsI(Na) to charged particles. Nucl. Instrum. Methods **68**(1), 9–12 (1969). http://www.sciencedirect.com/ science/article/pii/0029554X6990682X
6. T. Kim, I. Cho, D. Choi et al., Study of the internal background of CsI(Tl) crystal detectors for dark matter search. Nucl. Instrum. Methods Phys. Res. Sect. A **500**(1), 337–344 (2003). http:// www.sciencedirect.com/science/article/pii/S0168900203003462. NIMA Vol 500
7. N.E. Fields, CosI: development of a low threshold detector for the observation of coherent elastic neutrino-nucleus scattering. Ph.D. thesis, 2014. https://search.proquest.com/docview/ 1647745354?accountid=14657
8. Spallation Neutron Source Final Safety Assessment Document for Neutron Facilities (2011)
9. B. Scholz, CsI[Na] CEνNS analysis software package at GitHub (2017). https://doi.org/10. 5281/zenodo.918287

Chapter 6
Light Yield and Light Collection Uniformity

Given the large size of the CsI[Na] detector, a crucial calibration was to probe the light collection uniformity along the crystal. A ^{241}Am source was placed on the outside of the copper can at nine equally spaced positions along the length of the CsI[Na] detector (Fig. 6.1) to measure the light collection uniformity in the CsI[Na] crystal. The main γ-ray emission of this isotope only carries an energy of $E_\gamma = 59.54$ keV. This low energy ensures that interactions with the crystal occur in close vicinity to the source location. As a result, only a small volume of the crystal is irradiated for each source location. Comparing the total light yield, i.e., the number of SPE produced per unit energy, found at each position provides a measure for the nonuniformity in the detector response.

6.1 Detector Setup

The calibrations were performed at a sub-basement laboratory at the University of Chicago, providing roughly 6 m.w.e. of overburden as shielding from cosmic-rays. The central detector was placed within a well of contemporary lead, providing a total of 2 in. of γ-shielding on the bottom and 4 in. on the sides.

The detector was wired as shown in Fig. 5.3 with two exceptions. First, no muon veto was present during this calibration measurement. Second, the data was acquired triggering on the CsI[Na] signal itself instead of using an external trigger. The threshold of a standard edge trigger was set much lower than the typical amplitude of a ^{241}Am event. The threshold was further adjusted to achieve a data acquisition rate of $\mathcal{O}(60\,\text{Hz})$ to guarantee a 100% data throughput. The trigger position was set to 78.5% of the $70 - \mu s$-long traces, i.e., at sample 27,475. The CsI[Na] signal was sampled at 500 MS s^{-1} resulting in a total of 35,000 samples per waveform. The digitizer range was set to 200 mV$_{pp}$ with a digitizer depth of 8 bit. An example waveform is shown in Fig. 6.2. The CsI[Na] signal V

© Springer Nature Switzerland AG 2018
B. Scholz, *First Observation of Coherent Elastic Neutrino-Nucleus Scattering*,
Springer Theses, https://doi.org/10.1007/978-3-319-99747-6_6

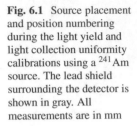

Fig. 6.1 Source placement and position numbering during the light yield and light collection uniformity calibrations using a ^{241}Am source. The lead shield surrounding the detector is shown in gray. All measurements are in mm

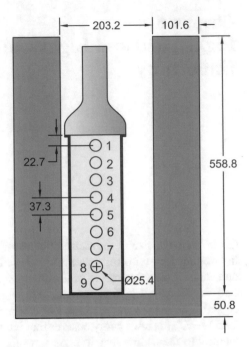

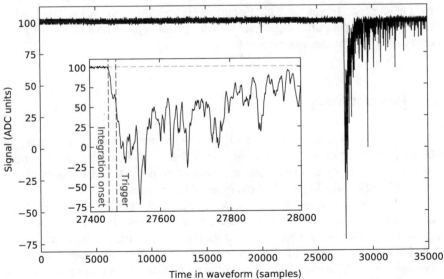

Fig. 6.2 Example waveform showing a typical ^{241}Am signal before data processing. The inset provides a zoom into the triggering region. Sample 27,400 marks the onset of the ROI which precedes the hardware trigger by 75 samples. Highlighted is the hardware trigger (dashed red) and the onset for the charge integration window (dashed blue)

was transformed using Eq. (5.2). For each source position, data was acquired for approximately 5 min.

6.2 Waveform Analysis

The analysis pipeline for each of the nine data sets was identical and is described in the following: Each acquired CsI[Na] waveform was divided into two distinct regions. The first is the pretrace (PT), which spans sample 0–27,399. The second region is the region of interest (ROI), spanning sample 27,400–34,999. The overall baseline was estimated using the median V_{median} of the first 20,000 samples ($=$ 40 μs) of the waveform. The CsI[Na] signal was shifted and inverted using

$$\hat{V}_i = V_{median} - V_i \quad \text{for} \quad i \in [0, 35,000).$$

(6.1)

All peaks within the waveform were detected using a standard threshold crossing algorithm [6]. A peak was defined as at least four consecutive samples showing a minimum deviation of at least 3 ADC counts from the baseline. The time of both positive and negative threshold crossings were recorded for each peak.

6.2.1 Mean SPE Charge Calibration

The SPE charge spectrum can be calculated using the peaks found in the pretrace (PT). The charge of each peak was determined by integrating the signal $\hat{V}$ within an integration window, which ranges from two samples before the positive threshold crossing of a peak to two samples after the negative threshold crossing.

The left panel of Fig. 6.3 illustrates the peak-finding algorithm and the definition of the integration window. A new baseline V^j_{median} was calculated for each peak j using the median of the 250 samples directly preceding and following the integration window. The total integrated charge for peak j was calculated using:

$$Q^j = \sum_{i=t_s}^{t_f} \left(\hat{V}_i - V^j_{median} \right),$$

(6.2)

where t_s denotes the index of the beginning of the integration window and t_f its end. The total charge Q^j is given in unconventional units of ADC counts $\times$ 2 ns. These units can be converted into conventional charge units using:

$$Q\,(\text{pC}) = \frac{Q\,(\text{ADC counts} \times 2\,\text{ns})}{32}$$

(6.3)

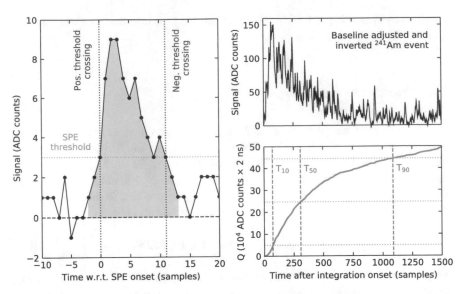

Fig. 6.3 **Left**: Process used to determine the integration window of an SPE. The signal was already baseline-adjusted and inverted. The dashed gray line represents the detection threshold. A peak is detected when at least four consecutive samples show at least a value equal to the detection threshold. The positive (red) and negative (blue) threshold crossings are recorded. The integration window ranges from two samples before the positive crossing to two samples after the negative crossing (shaded blue area). **Right**: The top panel shows an example of a baseline-adjusted and inverted ^{241}Am event. The onset was calculated using a standard threshold-finding algorithm. Shown is the full 1500 sample long integration window. The bottom panel shows the corresponding integrated charge in blue. The dotted gray lines represent the 10, 50, and 90% levels of the total charge integrated for this event. The corresponding timings for each of these threshold crossings, i.e., T_{10}, T_{50}, and T_{90}, are shown in orange

$$\text{as}\quad 1\,\text{ADC counts} \times 2\,\text{ns} \,\widehat{=}\, \frac{0.2\,\text{mV}}{2^8 \times 50\,\Omega} \times 2\,\text{ns} = 3.125 \times 10^{-14}\,\text{C},$$

However, no additional information was gained by its conversion and it was therefore omitted throughout the rest of this thesis.

Figure 6.5 shows the SPE charge spectrum calculated for data taken with the ^{241}Am source located at position two (Fig. 6.1). The distribution consists of all integrated peaks found in the PT region for all waveforms of this data set. The mean SPE charge Q_{spe} was calculated by fitting an SPE charge distribution model to the data. Several different models were proposed in the past [1, 7], and the average Q_{spe} obtained by fitting these models to the data can slightly differ. In the following paragraph, two competing models for the SPE charge distribution are introduced. The first uses a Gaussian distribution to describe the SPE charge spectrum, whereas the second uses a Polya distribution.

Most commonly, the shape of the SPE charge spectrum is approximated by a Gaussian distribution [1]. The corresponding fit function is given by:

$$f_{\text{gauss}}\left(q, a_n, \sigma_n, \vec{a}, Q_{\text{spe}}, \sigma, k, q_0\right) = \left[a_n e^{-q/\sigma_n} + \sum_{i=1}^{3} a_i \, g_i(q, Q_{\text{spe}}, \sigma) \right] \left(1.0 + e^{-k(q-q_0)}\right)^{-1}$$

$$\text{where} \quad g_i(q, Q_{\text{spe}}, \sigma) = \text{Exp}\left[\frac{(q - i \, Q_{\text{spe}})^2}{2 i \sigma^2} \right]. \tag{6.4}$$

Equation (6.4) includes the following components: The exponential e^{-q/σ_n} represents the events caused by random baseline fluctuations that exceed the SPE search criteria. g_i describes the i-th convolution of a Gaussian distribution with a mean of Q_{spe} and a standard deviation of σ with itself. The sum over g_i therefore includes the charge distributions for one ($i = 1$), two ($i = 2$), and three ($i = 3$) SPE. The multi-SPE peak amplitudes $a_{2,3}$ are allowed to float freely. The amplitudes $a_{1,2,3}$ are not restricted to follow a Poisson distribution, since the measurements of the SPE mean charge discussed throughout this thesis were not performed with a dedicated setup using an LED or laser pulse as trigger. The last factor in Eq. (6.4) describes a sigmoid-shaped acceptance curve, which represents the efficiency of the peak-finding algorithm to detect a peak of charge q. In this acceptance model, k describes the steepness of the sigmoid and q_0 its midpoint.

For some PMTs, the complex secondary electron emission at different dynode stages can lead to inconsistencies between the recorded experimental charge spectrum and a fitted model based on a Gaussian distribution. Prescott [7] instead proposed to use a Polya distribution to describe the SPE charge spectrum. Modifying Eq. (6.4) yields

$$f_{\text{polya}}\left(q, a_n, \sigma_n, \vec{a}, Q_{\text{spe}}, \sigma, k, q_0\right) = \left[a_n e^{-q/\sigma_n} + \sum_{i=1}^{3} a_i \, p_i(q, Q_{\text{spe}}, \sigma) \right] \left(1.0 + e^{-k(q-q_0)}\right)^{-1}$$

$$\text{where} \quad p_i(q, Q_{\text{spe}}, \sigma) = \text{Exp}\left[-(\sigma + 1) \frac{q}{Q_{\text{spe}}} \right] \left(\frac{q}{Q_{\text{spe}}} \right)^{i(\sigma+1)-1}, \tag{6.5}$$

where all Gaussian distributions g_i were replaced with Polya distributions p_i. Similar to the Gaussian case, the subscript i denotes the i-th convolution of a Polya distribution with itself. The noise contribution and the sigmoid-shaped acceptance curve remain unchanged.

Both Eqs. (6.4) and (6.5) were fitted to an SPE charge spectrum that was taken for an identical R877-100 PMT using an LED triggered setup, which was adapted from [5]. The experimental charge spectrum including both fits is shown in the top panel of Fig. 6.4. The bottom panel shows the deviation of each model from the data. Both distributions describe the peak region equally well. However, the overall deviation for the model based on the Polya distribution is smaller. For the remainder

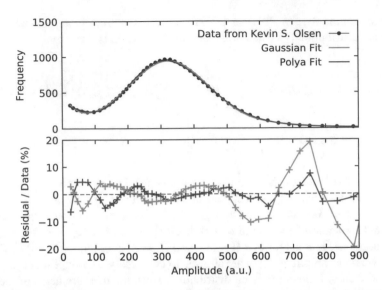

Fig. 6.4 Top: The data shown represents a dedicated SPE charge measurement for a R877-100 PMT using a LED triggered setup, which was taken from [5]. The SPE charge distribution was fitted using the models described in Eq. (6.4) (Gaussian, shown in blue) and (6.5) (Polya, shown in red). **Bottom**: Deviation between experimental data and the SPE charge model. Both models describe the peak region equally well. However, the overall deviation for the model based on the Polya distribution is smaller for the same degrees of freedom

of this thesis, a Polya-based SPE charge distribution model is used whenever an SPE charge distribution is fitted for the R877-100 PMT.

The mean SPE charge Q_{spe} was determined for each [241]Am source location by fitting Eq. (6.5) to the respective SPE charge spectra. Figure 6.5 shows the SPE charge spectrum calculated for the data acquired for source position two (Fig. 6.1). The best fit of Eq. (6.5) is shown in red. The individual fit components are shown in green (noise), blue (SPE charges), and orange (acceptance). The orange curve indicates that only a tiny fraction of low-charge SPEs are actually missed by the peak-finding algorithm. Comparing the counts obtained from integrating the fitted SPE charge model without the inclusion of the acceptance sigmoid to the counts obtained when including it shows that only $\mathcal{O}(0.5\%)$ of the SPEs were missed by the peak-finding algorithm. These SPEs further carry only a small charge and as a result only a negligible amount of total charge ($\ll 0.1\%$) remained unidentified.

The mean SPE charges Q_{spe} for all [241]Am source positions are shown in Fig. 6.6. A slight increase in Q_{spe} is apparent for [241]Am source distances of ≥ 10 cm. This suggests that the PMT was not completely warmed up during the data taking for the closest positions, resulting in a slightly smaller Q_{spe}. After several minutes, the amplification of the PMT stabilized and resulted in a charge variation of less than

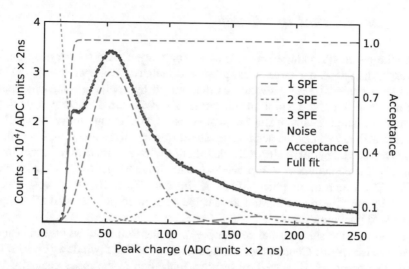

Fig. 6.5 SPE charge distribution as determined for ^{241}Am source position two (Fig. 6.1). Shown in red is the best fit of Eq. (6.5) to the data. All individual fit components are also shown. Even though a Polya distribution was used as the SPE charge model, only a small deviation from a normal distribution is visible for the blue curves. The SPE charge distribution for all other ^{241}Am source positions showed similar fit qualities

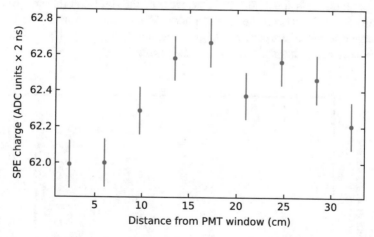

Fig. 6.6 Mean SPE charge calculated for all ^{241}Am source positions. Only a 1% variation in SPE charge is visible among the different runs

1%. This does not affect the light yield measurements in any significant way. The trigger threshold was chosen well below the amplitude expected from a 59.54 keV energy deposition. As a result, all ^{241}Am events trigger the DAQ as a variation in their amplitude of ~1% is too small to introduce any bias.

6.2.2 Arrival Time Distribution

Once the mean SPE charge was calculated for every ^{241}Am source position, the corresponding ^{241}Am charge spectra were calculated. For each waveform, the onset of a potential ^{241}Am event was determined by applying a standard peak-finding algorithm to the region of interest (ROI). Peaks are defined by at least ten consecutive samples with an amplitude of at least 3 ADC counts above the baseline. The arrival time T_{arr}, i.e., the positive threshold crossing of the first peak found in the ROI is recorded. Figure 6.7 shows the distribution of all recorded arrival times for all data sets. The beginning of ROI precedes the hardware trigger by 75 samples (inset of Fig. 6.2), which limits potential arrival times to $T_{arr} \in [0, 75]$ sample. Events caused by the ^{241}Am γ emission are visible as a broad peak around 55 samples after the ROI window onset.

In contrast, the sharp rise at 69 samples after the ROI window onset is caused by afterpulses [4] or Cherenkov light emission in the PMT window [9]. The first of the two processes is caused by residual ionization in the gases caused by the accelerated photoelectrons inside the PMT. The ions produced in this ionization are accelerated towards the photocathode. Upon impact on the photocathode, an afterpulse is generated [4]. The second is caused by small trace amounts of ^{238}U, ^{232}Th, or ^{40}K within the PMT window. The β-decays of such isotopes can produce Cherenkov light emission in the PMT window or the electrons released in such a decay directly interact with the photocathode. The corresponding signal is a tight pulse large enough to surpass the trigger threshold (Fig. 6.8). These Cherenkov events usually carry charges equivalent to 2–15 SPE [2, 9].

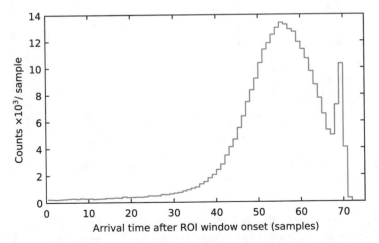

Fig. 6.7 Arrival time distribution of events triggering the data acquisition system in the ^{241}Am calibrations. A clear peak is visible at approximately 55 samples, arising from ^{241}Am events. As a result, it takes one of these scintillation light curves approximately 20 samples to rise from the baseline up to the trigger threshold. The sharp rise at 69 samples is caused by Cherenkov pulses, which are described in the text. These events show a much sharper rising edge

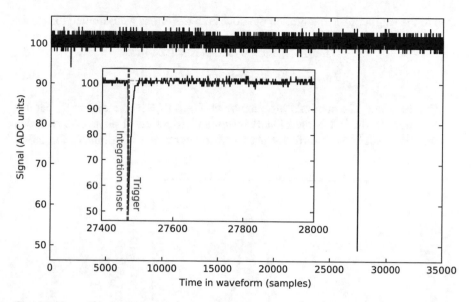

Fig. 6.8 Example waveform showing an event triggering on a Cherenkov signal arising from Cherenkov light emission in the PMT window. The large spike is typical of such an event and contains a charge equivalent to multiple SPE. The inset shows the hardware trigger (dashed red) and the onset for the integration window as determined by the peak-finding algorithm (dashed blue). It is apparent that for this type of background the hardware trigger and the peak onset determined by the peak-finding algorithm almost coincide, leading to an arrival time T_{arr} of $\sim$70 samples

6.2.3 Rise-Time Distributions

To calculate the total charge for an event in ROI, a $3 - \mu s$-long integration window was defined. The onset of this window is given by two samples before T_{arr} and spans a total of 1500 samples, i.e., $3\,\mu s$. A new baseline $V_{baseline}$ was estimated for this integration window using the median of the 500 samples ($1\,\mu s$) directly preceding it. The integrated scintillation curve $Q(t)$ is given by:

$$Q(t) = \sum_{i=T_{arr}}^{T_{arr}+t} \left[\hat{V}_i - V_{baseline} \right]$$ (6.6)

For each event, the total integrated charge $Q_{total} = Q(1499\,S)$ was recorded. In addition, two different rise-times [3, 8] were calculated using $Q(t)$. Rise-times are defined as the time it takes the integrated scintillation curve $Q(t)$ to rise from one predefined percentage of its maximum to another. The right panels of Fig. 6.3 provide an illustration of their calculation.

The first rise-time was $T_{0-50} = T_{50}$, i.e., the time it took the integrated charge to rise from 0 to 50% of the total charge integrated in the window. The second was $T_{10-90} = T_{90} - T_{10}$, i.e., the time between passing the 10 and 90% charge threshold.

Due to the length of the integration window, both rise-times were limited to:

$$T_{0-50} \in [0, 1500) \text{ samples} = [0, 3)\mu s \tag{6.7}$$

$$T_{10-90} \in [0, 1500) \text{ samples} = [0, 3)\mu s. \tag{6.8}$$

The rise-time distributions are shown in Fig. 6.9. Several different features are apparent in the two-dimensional rise-time distribution of all events recorded (bottom-left panel). Four different phase-space regions are highlighted. Circled in

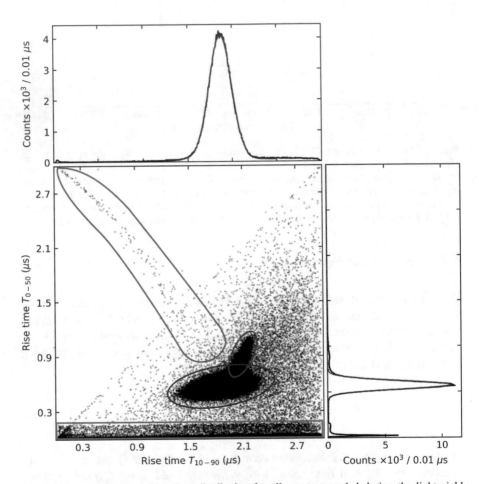

Fig. 6.9 Two-dimensional Rise-time distribution for all events recorded during the light yield calibration. The integration window was $3\,\mu s$ long. Four distinct features are apparent. **Red**: Proper radiation-induced events, mainly from γs with an energy of $E = 59.54\,\text{keV}$. **Blue**: Events exhibiting a digitizer range overflow. For these events, the first couple of nanoseconds were clipped by the digitizer, which lead to longer rise-times. **Orange**: The pulse onset was misidentified due to a spurious SPE preceding the pulse. **Green**: Triggers on Cherenkov pulses with and without a leading SPE

Table 6.1 Mean and standard deviation of a Gaussian fitted to the marginalized rise-time distributions of T_{0-50} and T_{10-90}

	T_{0-50}		T_{10-90}
μ_{0-50}	$0.57430 \pm 0.00048\,\mu s$	μ_{10-90}	$1.86958 \pm 0.00059\,\mu s$
σ_{0-50}	$0.04618 \pm 0.00037\,\mu s$	σ_{10-90}	$0.12726 \pm 0.00050\,\mu s$

The fits are shown in Fig. 6.9

red are correctly identified radiation-induced events, mainly from ^{241}Am γ emission (Fig. 6.2). These events are centered at $T_{10-90} = 1.87\,\mu s$ and $T_{0-50} = 0.57\,\mu s$. It can be observed from the marginalized rise-time distributions on the top and right panel of Fig. 6.9 that most events recorded fell within this category. A Gaussian was fitted to both marginalized rise-time distributions. Even though the ellipse covering these events appears tilted in the two-dimensional distribution, an excellent agreement between data and fit was found. The fit results are given in Table 6.1.

Circled in blue are events that exceeded the digitizer range. Both rise-times were biased towards higher values as the respective event signal was truncated in the early part of the integration window. The arrival time T_{arr} of events circled in orange was misidentified due to a preceding spurious peak in the ROI, leading to a misaligned integration window. Last, a fourth population consisting of triggers on a Cherenkov pulse (Fig. 6.8) are circled in green. Their spread is caused by spurious peaks before and after the triggering Cherenkov spike.

6.3   Determining the Light Yield and Light Collection Uniformity

The information provided above was used to reject background events. First, all events were rejected which showed at least one sample exceeding the digitizer range. Next, the following cuts on rise-time were implemented to remove Cherenkov triggers and events with a misidentified onset:

$$T_{0-50} > 0.25\,\mu s = 125\ \text{samples} \tag{6.9}$$

$$T_{10-90} > T_{0-50} \tag{6.10}$$

These rise-time cuts were restrictive enough to remove most background events without rejecting any ^{241}Am events. In addition, events with more than $N_{pt}^{max} = 3$ peaks in the PT were rejected to minimize any bias from afterglow from a preceding event. The charge of all remaining events was converted into an equivalent number of SPE N_{pe} using the corresponding mean SPE charge Q_{spe}. It is

$$N_{pe} = \frac{Q_{total}}{Q_{spe}} \tag{6.11}$$

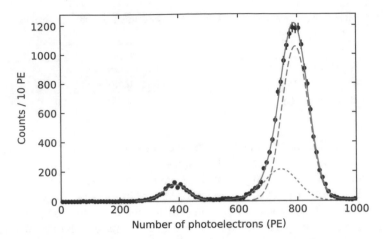

Fig. 6.10 Energy spectrum recorded for the ^{241}Am source position two. The integrated charges were converted to a corresponding number of PE using Q_{spe}. The main ^{241}Am emission peak is apparent (dashed), which also contains contributions from the L-shell escape peak of cesium and iodine (dotted). A smaller peak, which is visible at lower energies, represents the corresponding K-shell escape peak

The energy spectrum for ^{241}Am source position two (Fig. 6.1) is shown in Fig. 6.10. The energy spectra for all other source locations were comparable. The main ^{241}Am emission peak is apparent at $\sim$800 PE. The shoulder towards lower energies is caused by the L-shell escape peaks of cesium and iodine. A secondary feature at $\sim$400 PE represents the K-shell escape peaks from cesium and iodine. Due to their low energy, γ-rays from ^{241}Am interact with the CsI[Na] crystal close to the surface making such escapes likely.

For each source position, the average number of photoelectrons N_{pe}^{am} produced in a ^{241}Am event was determined from the respective energy spectra. Using N_{pe}^{am}, the local light yield $\mathcal{L}$ for an incoming energy of 59.54 keV was calculated by:

$$\mathcal{L} = \frac{N_{pe}^{am}}{59.54\,keV} \tag{6.12}$$

Figure 6.11 shows the light yield calculated for every source position. Only a small variation in $\mathcal{L}$ is apparent for all source locations, which was expected from discussions with the manufacturer. The largest deviation was found at the position farthest away from the PMT window and closest to the back reflector. The uncertainty weighted average of the closest eight positions is given by:

$$\mathcal{L}_{CsI} = 13.348 \pm 0.019\,\frac{PE}{keV_{ee}}. \tag{6.13}$$

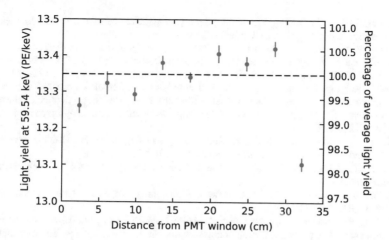

Fig. 6.11 Light yield for every source position. The eight positions closest to the PMT show a variation in light yield of $\mathcal{O}(0.5\%)$. The largest deviation was found for the source position farthest from the PMT and closest to the backing reflector. A dotted line shows the weighted average of the closest eight positions at $\mathcal{L}_{CsI} = 13.348\,\mathrm{PE/keV_{ee}}$

Table 6.2 Variation of the light yield at 59.54 keV along the crystal axis

Distance from PMT (cm)	Light yield (PE/keV$_{ee}$)	Variation from $\mathcal{L}_{CsI}$ (%)
2.3	13.261 ± 0.021	−0.65
6.0	13.324 ± 0.032	−0.19
9.7	13.293 ± 0.018	−0.41
13.5	13.380 ± 0.019	0.24
17.2	13.341 ± 0.014	−0.059
20.9	13.406 ± 0.024	0.43
24.7	13.379 ± 0.020	0.23
28.4	13.421 ± 0.020	0.54
32.1	13.104 ± 0.018	−1.8

$\mathcal{L}_{CsI}$ was found to be 13.348 ± 0.019 (PE/keV$_{ee}$)

The overall variation with respect to $\mathcal{L}_{CsI}$ within the first eight locations is $\mathcal{O}(0.5\%)$. The position farthest away shows a deviation of 1.8%. The light yield values and their respective deviation from $\mathcal{L}_{CsI}$ for each position are given in Table 6.2. This measurement confirmed an almost perfect light collection uniformity throughout the crystal. For the rest of this thesis, a constant light yield given by $\mathcal{L}_{CsI}$ was assumed.

References

1. E. Bellamy, G. Bellettini, J. Budagov et al.,   Absolute calibration and monitoring of a spectrometric channel using a photomultiplier. Nucl. Instrum. Methods Phys. Res. Sect. A Accel. Spectrom. Detect. Assoc. Equip. **339**(3), 468–476 (1994). http://www.sciencedirect.com/science/article/pii/016890029490183X

2. E. Enterprises. Understanding Photomultipliers. Online (2011). http://www.et-enterprises.com/files/file/Understanding-photomultipliers.pdf
3. X. Luo, V. Modamio, J. Nyberg et al., Test of digital neutron–gamma discrimination with four different photomultiplier tubes for the NEutron Detector Array (NEDA). Nucl. Instrum. Methods Phys. Res. Sect. A Accel. Spectrom. Detect. Assoc. Equip. **767**, 83–91 (2014)
4. K. Ma, W. Kang, J. Ahn et al., Time and amplitude of afterpulse measured with a large size photomultiplier tube. Nucl. Instrum. Methods Phys. Res. Sect. A Accel. Spectrom. Detect. Assoc. Equip. **629**(1), 93–100 (2011). http://www.sciencedirect.com/science/article/pii/S0168900210026306
5. K.S. Olsen, Improvements to the resolution and efficiency of the DEAP-3600 dark matter detector and their effects on background studies (2010). Master's thesis
6. Peak Detection Using LabVIEW and Measurement Studio (2017). http://www.ni.com/white-paper/3770/en/
7. J.R. Prescott, A statistical model for photomultiplier single-electron statistics. Nucl. Instrum. Methods **39**(1), 173–179 (1966)
8. E. Ronchi, P.-A. Söderström, J. Nyberg et al., An artificial neural network based neutron–gamma discrimination and pile-up rejection framework for the BC-501 liquid scintillation detector. Nucl. Instrum. Methods Phys. Res. Sect. A Accel. Spectrom. Detect. Assoc. Equip. **610**(2), 534–539 (2009)
9. S. Roodbergen, R. Kroondijk, H. Verheul, Cherenkov radiation in photomultiplier windows and the resulting time shift in delayed coincidence time spectra. Nucl. Instrum. Methods **105**(3), 551 – 555 (1972). http://www.sciencedirect.com/science/article/pii/0029554X72903539

Chapter 7
Barium Calibration of the CEνNS Detector

After establishing that the variation of light collection efficiency along the CsI[Na] crystal axis was negligible, an additional low-energy calibration was performed. The goal of this measurement was to build a library of low-energy events taking place within the CsI[Na] crystal that can be used to define and quantify the acceptance of several different cuts used later in the analysis of the CEνNS search data. Given the minimal difference in the decay times for nuclear and electronic recoils in CsI[Na] at low energies (Fig. 5.6), this calibration was performed using electronic recoils induced by γ interactions instead of relying on a neutron source to induce nuclear recoils. The CsI[Na] detector used for these calibrations was later used for the CEνNS search (Chap. 9).

The total data acquisition for this calibration took place over the course of three months, i.e., between March 27, 2015 and June 15, 2015. This measurement provided an ideal test environment for the long term stability of the detector setup. No issues were identified and as a result the detector was deployed at the SNS shortly after the calibration. The detector setup was located at the same sub-basement laboratory where the light yield calibrations described in Chap. 6 were performed. In this location, the CsI[Na] detector was shielded against activation from cosmic rays for a prolonged time by $\sim$6 m.w.e., reducing the backgrounds present in the CEνNS search.

7.1 Detector Setup

In order to create an event library containing interactions with only a few to a few tens of SPE within the crystal, a highly collimated pencil beam of ^{133}Ba γ-rays was sent through the CsI[Na]. This pencil beam was produced by placing a ^{133}Ba button source in a lead collimator. The lead collimator provided a pinhole with a diameter of $\varnothing = 1.2$ mm and a total depth of $d = 44.5$ mm. This constrained

© Springer Nature Switzerland AG 2018
B. Scholz, *First Observation of Coherent Elastic Neutrino-Nucleus Scattering*,
Springer Theses, https://doi.org/10.1007/978-3-319-99747-6_7

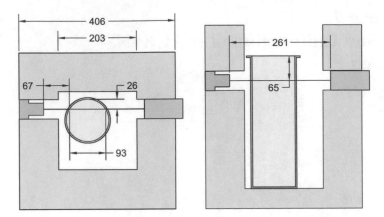

Fig. 7.1 Schematic of the detectors and shielding used in the ^{133}Ba calibration measurement. Measurements are in mm. A top-down view is shown on the left, and a side view is provided on the right. The PMT and the lead shield on top were omitted for clarity. The red object represents the lead collimator used to constrain the isotropic emission of the ^{133}Ba button source to a pencil beam. The blue object represents the BrilLanCe^{TM} backing detector. The pencil beam only traverses the edge of the crystal at a given height. However, as established in Chap. 6, the variation in the light collection efficiency along the crystal is negligible. As a result, this calibration is representative of events happening anywhere in the CsI[Na] crystal

the maximum aperture of the collimated ^{133}Ba γ-beam to $\theta_c \sim 1.9°$. The setup triggered on forward Compton-scattered gammas detected using a BrilLanCe^{TM} backing detector. A schematic of the experimental setup is shown in Fig. 7.1. Pictures of the setup are shown in Fig. 7.2. The maximum single forward scattering angle of a γ-ray being detected in the backing detector is given by $\theta_{max} \sim 12°$, leading to a maximum energy deposition of $E_{max} \sim 6\,keV$.

The schematic of the data acquisition system used in this measurement is shown in Fig. 7.3. The linear-gate logic rejecting high-energy events in the CsI[Na] signal is identical to the one described in Chap. 5. However, a different triggering signal was used to acquire data. The pre-amp output of the BrilLanCe^{TM} backing detector was fed into an Ortec 550 single-channel analyzer (SCA). This module provided a logical signal, acting as external trigger, whenever an event with an equivalent energy of 200–500 keV$_{ee}$ was detected within the BrilLanCe^{TM} backing detector. This ensured that a trigger was active for all small angle Compton-scattered gammas of all major emission lines of ^{133}Ba. For each trigger, waveforms with a total length of 35,000 samples at a sampling rate of 500 MS s^{-1}, i.e., 70 μs, were recorded. In contrast to the previous calibration, the trigger location was set to 100% of the trace, i.e., at 70 μs into the trace. This is necessary to account for the jitter introduced by using the SCA. The SCA produces a trigger whenever the analyzed signal exits the analysis window (Fig. 7.4). This behavior causes the timing of the trigger produced by the SCA to be directly dependent on the energy deposited in the backing detector. Consequently, events with higher energy show larger lag times between the CsI[Na] signal and the trigger from the backing detector (Fig. 7.4).

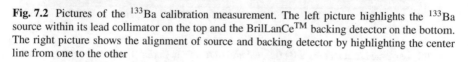

Fig. 7.2 Pictures of the ^{133}Ba calibration measurement. The left picture highlights the ^{133}Ba source within its lead collimator on the top and the BrilLanCe^{TM} backing detector on the bottom. The right picture shows the alignment of source and backing detector by highlighting the center line from one to the other

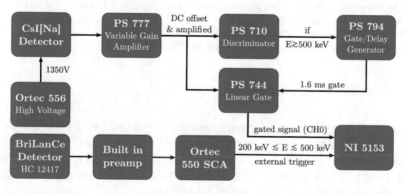

Fig. 7.3 Data acquisition system used in the ^{133}Ba calibration measurement

While adjusting the SCA analysis levels, the CsI[Na] output was closely monitored to ensure that the maximum delay, i.e., $T_{lag} + \Delta T$, between a trigger from the SCA and a scattering event in the CsI[Na] was less than 20 μs. Therefore, all Compton-scattered events occur no sooner than 50 μs into the waveform. However, not all triggers by the backing detector are due to small angle Compton-scattered events. Even though the BrilLanCe^{TM} detector was surrounded by some lead, some triggers were caused by environmental radiation that deposited the right energy within the backing detector. As the triggering particle did not interact with the CsI[Na] prior to hitting the backing detector, the corresponding waveforms of these events only consist of random coincidences between the CsI[Na] and the backing detector.

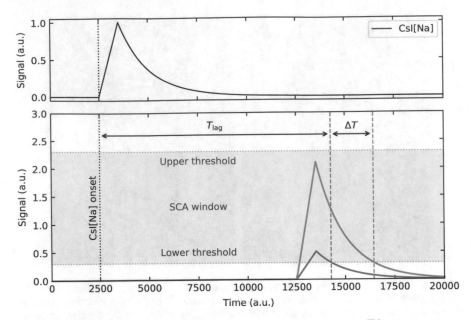

Fig. 7.4 Lag time between the CsI[Na] signal and trigger from the BrilLanCe^{TM} backing detector. The CsI[Na] signal is shown on the top, whereas the corresponding BrilLanCe^{TM} signal is shown on the bottom. The red (blue) curve represents an event depositing an energy close to $\sim$200 keV$_{ee}$ ($\sim$500 keV$_{ee}$) in the backing detector. The difference in the actual trigger position of both events, i.e., when a signal actually leaves the SCA window, is readily visible

7.2   Waveform Analysis

As the low-energy events of interest consist of only a few PE, it is challenging to determine whether a singular event was acquired due to a true Compton-scattered trigger or due to environmental background radiation. However, true and random coincidences can be discriminated on a statistical basis by using all 21,275,438 waveforms acquired over the course of the 3 months.

To achieve this statistical discrimination, two slightly overlapping regions are defined in each waveform (Fig. 7.5). The first interval is the anti-coincidence (AC) region that extends from sample 150 to 25,000. As discussed earlier, it is impossible for a true coincidence event to occur before sample 25,000 (= 50 μs into the waveform). This immediately implies that this region can only contain random coincidences between events in the CsI[Na] and the backing detector. The second interval denotes the coincidence (C) region, extending from sample 7650 to 32,500. This region contains both, random coincidences from environmental triggers and true coincidence events from small angle Compton scattering. Both regions are analyzed using an identical analysis pipeline. Through this process, two independent n-tuples are created for each waveform, one for the C and one for the AC region. These n-tuples contain parameters characterizing each event, e.g., the

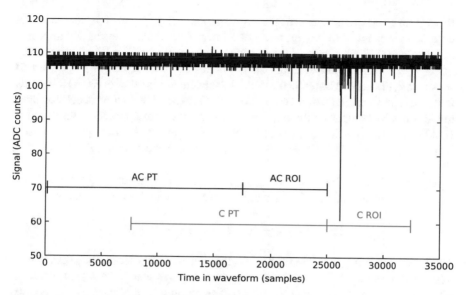

Fig. 7.5 Example waveform for the ^{133}Ba calibration highlighting different analysis regions. The trigger from the Ortec 550 SCA was set to sample 35,000. No true coincidence event can occur prior to sample 25,000. The two main analysis regions are also shown. The anti-coincidence region is depicted in black and the coincidence region in red. Their respective sub-regions are also included, i.e., pretrace (PT) and region of interest (ROI)

Table 7.1 Start and end sample of each analysis region

	AC region		C region	
	Start	End	Start	End
PT	150	17,500	7650	25,000
ROI	17,500	25,000	25,000	32,500

The starting sample is included in the respective region, whereas the ending sample is excluded. Here, PT denotes the pretrace and ROI the region of interest

average baseline or the number of peaks present in an analysis region (Table 7.2). These two distinct sets of n-tuples are denoted as C and AC data sets throughout the remainder of this chapter (Table 7.1).

To extract a certain feature exhibited by low-energy events, the statistics of the feature can be extracted by subtracting its distribution derived from AC from the distribution derived from C. The residual $\mathcal{R}=C\text{-}AC$ only contains a statistical contribution from low-energy events only, as long as both regions were treated identically throughout the full analysis. The analysis pipeline for each waveform is detailed below.

For each waveform, it is determined whether the signal is fully contained within the digitizer range. A digitizer overflow is determined as any sample that shows as $+127$ or -128 ADC counts, i.e., the upper and lower limits of an I8 variable.

If such an overflow is detected, the corresponding flag $f_o = 1$ is set. Next, the waveform is checked for the presence of a linear gate. Such an event will appear as a baseline of 0 ADC counts, as the CsI[Na] signal is blocked for 1.6 ms whenever a linear gate is initiated due to a high-energy deposition within the CsI[Na] (Chap. 5). In contrast, the normal baseline for the CsI[Na] detector is set to ∼100 ADC counts. It is possible to implement a computationally inexpensive way to check for these linear gates by comparing the number of times a waveform crosses a threshold of 18 ADC counts in a falling (n_f) and rising (n_r) manner. A linear gate is present if $n_f \neq n_r$, which is indicated in the data by the corresponding linear-gate flag $f_g = 1$.

To determine a global baseline V_{median}, the median of the first 20,000 samples of the current waveform is used. The signal V is adjusted and inverted, that is:

$$\hat{V}_i = V_{median} - V_i \quad \text{for} \quad i \in [0, 35,000). \tag{7.1}$$

The location of any potential SPE can be identified using a peak-finding algorithm as described in the light yield calibration (Sect. 6.2.1). As stated, a peak was detected when at least four consecutive samples had an amplitude of at least 3 ADC counts. Both positive and negative threshold crossings were recorded for each peak. Similar to the procedure in Sect. 6.2.1, the charge of each peak is calculated using Eq. (6.2).

Instead of calculating a single, overall SPE charge distribution for the full data set, the data is subdivided into 5-min intervals to monitor the stability of the mean SPE charge Q_{spe}. For each interval, an independent charge distribution is created. The distributions are fitted using Eq. (6.5) to extract the mean SPE charge Q_{spe} for the corresponding data period. In the further data analysis, the appropriate Q_{spe} is used to establish a proper energy scale.

In what follows, the procedure applied to the analysis regions, i.e., C and AC, is described. The analysis is identical for both regions and illustrated in Fig. 7.6. A full example waveform is shown in panel **a**. The C (AC) region is shown in shaded red (gray). A zoom into both regions is shown in panel **b** (AC) and **c** (C), respectively. The nonoverlapping PT and ROI sub-regions are highlighted. The pretrace (PT) is fairly long, spanning 17,350 samples. Its position was chosen such that it could not possibly contain a signal from a coincident Compton-scatter event for neither analysis region. Its main purpose is to provide a veto against contamination from afterglow. As discussed in Chap. 5, and shown in Fig. 5.7, CsI[Na] can phosphorescence for up to $\mathcal{O}(10\,ms)$ after an actual event. A large energy deposition, e.g., from a muon traversing the crystal, could potentially add several SPEs from phosphorescence in a subsequent trigger and introduce a nonnegligible bias in the analysis. To remove these potentially contaminated events from the data set, events were rejected based on the total number of peaks detected in the corresponding PT. The higher the number of peaks, the likelier it is that these were caused by phosphorescence.

The region of interest (ROI) is much shorter, spanning 7500 samples. The AC ROI is aligned such that it cannot physically contain a coincident Compton-scatter event, whereas the C ROI could contain these events.

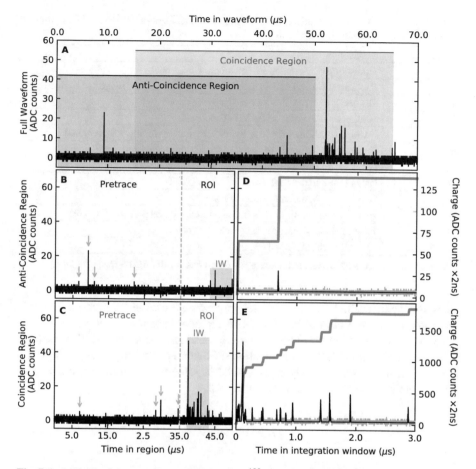

Fig. 7.6 Analysis of an example waveform for the ^{133}Ba calibration. The full waveform is shown in panel (**a**) with the full AC and C regions highlighted. Panels (**b**) and (**c**) provide a zoom into AC and C, respectively, with the PT and ROI sub-regions annotated. Green arrows mark all peaks in the respective PT regions as identified by the threshold-finding algorithm. The shaded blue region highlights the 3-μ s-long integration window, starting with the first peak in the ROI. Panels (**d**) and (**e**) provide a further zoom into these integration windows. The gray curve shows the baseline-adjusted signal. The black one in contrast represents the signal after zero-suppressing all samples that were not identified as part of a peak. The blue curve represents the charge integration $Q(t)$ of the zero-suppressed signal that is later used to determine the total charge as well as the individual rise-times

Peaks within both regions, i.e., PT and ROI, are identified using the results from the initial peak search. The total number of peaks N_{pt} in the PT is recorded. For the example waveform shown in Fig. 7.6, it is $N_{pt}^{C} = N_{pt}^{AC} = 4$ as indicated by the green arrows. The location of the first peak within the ROI defines the onset of the 1500 sample (=3 μs) long integration window. These windows are highlighted in shaded blue in panels **b** and **c**. A new baseline for each integration window is determined

Table 7.2 Overview of the parameters recorded for each waveform in the ^{241}Am calibration, the ^{133}Ba calibration, and the CEνNS search

Parameter	Description
t	Timestamp of current event in seconds since epoch
f_o	The waveform contains samples exceeding the digitizer range
f_g	The waveform contains a linear-gate event
V_{median}	Median baseline of the first 20,000 samples
V_{avg}	Average baseline of the first 20,000 samples (excluding peak regions)
σ_v	Standard deviation of the first 20,000 samples (excluding peak regions)
N_{pt}	Peaks found in the PT
N_{roi}	Peaks found in the ROI
N_{iw}	Peaks found in the integration window
T_{arr}	Arrival time, i.e., time of first peak in ROI with respect to ROI onset
Q_{total}	Total integrated charge in integration window
T_{10}, T_{50}, T_{90}	Rise-times of integrated scintillation curve

Two independent data sets are created, one for the C and one for the AC region

using the median of the 1 μs immediately preceding the window. A zoom into both integration windows after adjustment with the new baseline is shown in gray in panels **d** and **e**. Any sample not belonging to a peak is zero-suppressed. The resulting signal $V^\star$ is shown in black. It can be observed that all peaks are preserved, whereas the baseline noise is completely suppressed. The integrated scintillation curve $Q(t)$ (shown in blue) for both integration windows is

$$Q(t) = \sum_{i=T_{arr}}^{T_{arr}+t} V_i^\star \qquad (7.2)$$

Besides the total integrated charge $Q_{total} = Q(1499\,S)$ for each event, the rise-times T_{10}, T_{50}, and T_{90} are also recorded, as described in Chap. 6 and shown in Fig. 6.3.

The parameters extracted for each event are given in Table 7.2, where a separate and independent n-tuple is created for both the AC and C regions containing their respective parameters where applicable. The procedure described above is repeated for each waveform, and the individual n-tuples are gathered in the $\mathcal{AC}$ and $\mathcal{C}$ data sets.

7.3   Detector Stability Performance Over the Data-Taking Period

After all recorded waveforms were analyzed, the detector stability over the course of the data-taking period was tested. There are five main parameters of interest, which are all shown in Fig. 7.7. Each data point represents the average over a

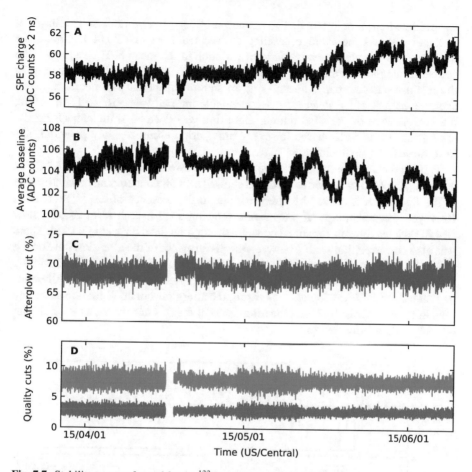

Fig. 7.7 Stability tests performed for the ^{133}Ba calibration. Every data point represents the average of a five-minute interval. Panel (**a**) shows the evolution of the mean SPE charge. Panel (**b**) shows the average baseline of the first 40 μs of each waveform. An anti-correlation between both parameters is evident. A change of 1 ADC counts in the average baseline corresponds to ~1 ADC counts × 2 ns, which could point towards a bias in the SPE charge calculation. However, the integration window of an SPE typically spans ~17 samples. A significant bias in the baseline calculation of the SPE charge integration window would therefore result in a change in the SPE charge on the order of $\mathcal{O}(17$ ADC counts × 2 ns). A change of ~1 ADC counts × 2 ns is therefore deemed acceptable. In addition, no increase or decrease in the average number of PE detected per five-minute interval was observed. Panel (**c**) shows the percentage of waveforms with more than ten peaks in the anti-coincidence pretrace. The high percentage of waveforms with such a large number of peaks in the pretrace can be attributed to the high level of phosphorescence in the crystal due to the presence of the ^{133}Ba source. No significant variation over time was seen in the percentage of waveforms rejected. No correlation with the mean SPE charge (panel **a**) or the average baseline (panel **b**) is apparent. The percentage of waveforms showing more than ten peaks in the coincidence pretrace shows the same level of phosphorescence and is therefore not depicted. The last panel **d** shows the percentage of waveforms showing either a linear gate (red) or digitizer overflow (blue), which are again stable throughout the whole data-taking period

five-minute-long time interval. The evolution of Q_{spe} over time is shown in panel **a**. Panel **b** shows the average baseline derived from the first 20,000 samples for the same five-minute intervals. A slight variation in the mean SPE charge over the course of the measurement that appears to be anti-correlated to the fluctuations in the average baseline can be seen. Yet, no significant increase or decrease in the average number of PE detected per five-minute interval was observed. Therefore, the performance of the SPE finding algorithm was deemed sufficient. Using the appropriate mean SPE charge for every time period eliminates any potential bias introduced by these fluctuations.

Panel **c** shows the number of waveforms with more than ten peaks in the AC PT. Almost, 70% of all waveforms recorded showed a significant amount of scintillation in the PT, which is caused by afterglow due to the presence of the ^{133}Ba source. However, the percentage of waveforms showing such a high level of afterglow within their waveforms remained constant throughout the data-taking period. There was also no correlation visible between the performance of the afterglow cut and the average baseline or the mean SPE charge. The distribution of peaks in the C and AC PT marginalized over the full data set is shown in Fig. 7.8. No significant difference between $\mathcal{AC}$ and $\mathcal{C}$ was found. As a result, the afterglow cut does not introduce any bias in the event selection, and the afterglow cut for $\mathcal{C}$ is not shown as it would not add any further information.

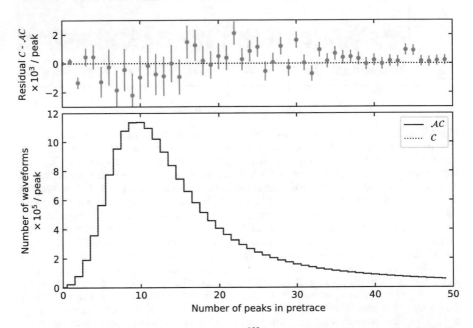

Fig. 7.8 Number of peaks in the pretrace for the ^{133}Ba calibration. The bottom panel shows the raw distributions for both $\mathcal{AC}$ (black) and $\mathcal{C}$ (red), whereas the top panel shows the residual $\mathcal{R}$ of both data sets. A slight deviation from zero is visible for low N_{pt}. However, the average difference between both distributions is only 0.24%, which is deemed negligible

Panel **d** of Fig. 7.7 shows the percentages of events with either a linear-gate event (red) or a digitizer range overflow (blue). Both showed no significant variation over the course of the data taking and are independent of any variation in other parameters. It is therefore assumed that the data quality cuts do not introduce any bias in the event selection.

In summary, the detector performed satisfactory over the full three months of data taking. No time periods from the data analysis need to be excluded and the full data set containing $\sim$21 million triggers could be used to quantify the cut acceptances.

7.4 Definition and Quantification of Data Cuts

After establishing that the CsI[Na] detector performed as expected throughout the full data-taking period of this ^{133}Ba calibration, the statistical data analysis was performed. As discussed above, separate n-tuples were created for each trigger and each analysis region. All n-tuples belonging to the AC region are denoted as $\mathcal{AC}$ data set and all n-tuples belonging to the C region are denoted as $\mathcal{C}$ data set. For both data sets, an additional entry was added to each n-tuple by converting the total charge integrated in the integration window into an equivalent number of PE N_{pe}. The conversion used the corresponding mean SPE charge Q_{spe} determined for the five-minute interval that contains the event, that is:

$$N_{pe} = \frac{Q_{total}}{Q_{spe}} \tag{7.3}$$

First, a series of data quality cuts were applied to both $\mathcal{AC}$ and $\mathcal{C}$. All events containing a linear gate f_g and/or overflow flag f_o were removed from analysis. As these flags are raised for the full waveform, they are shared between both sets, and the events removed were identical for $\mathcal{AC}$ and $\mathcal{C}$. Second, all events that showed more than N_{pt}^{max} peaks in their respective PT were removed from each data set. The purpose of this data cut is to minimize any bias introduced by excessive afterglow from previous energy depositions. Given the slight difference in the onset of the AC and C PT, an individual event can be deemed good in one data set, whereas it was cut in the other. However, the overall number of events passing this cut is neither biased towards $\mathcal{AC}$ nor $\mathcal{C}$ (Fig. 7.8). The total number of events cut is also dependent on the exact choice of N_{pt}^{max} and is further examined later on. This concludes the discussion of quality cuts for the barium calibration. In the following, the acceptance calculations for the data cuts used in the CEνNS search are presented.

7.4.1 The Cherenkov Cut

The first data cut that was applied is the so-called *Cherenkov cut*. As discussed earlier, a single Cherenkov pulse usually carries a charge equivalent to 2–15 SPE

(Chap. 6), and as such would pose an unwanted background to the CEνNS search. However, such an event typically only produces a single, sharp spike in the digitizer trace, whereas a CEνNS event appears as multiple individual SPE peaks following a scintillation decay time as shown in Fig. 5.6. The cut is therefore set to allow only events where the total number of peaks found in the integration window is at least N_{iw}^{min}. This cut was applied to events with a total energy of $N_{pe} \leq 40$ to avoid high-energy events from being rejected by the Cherenkov cut. With increasing N_{pe}, i.e., increasing energy, individual SPE will merge into larger peaks. As the peak-finding algorithm only tags peaks by their threshold crossings, this would lead to a decreasing number of peaks found. Applying the Cherenkov cut only to data with $N_{pe} \leq 40$ guarantees that no high-energy event was mistakenly rejected by this cut.

Figure 7.8 shows that the average number of peaks found in each PT, i.e., AC and C region of triggers in the ^{133}Ba calibration, is on the order of ten. Assuming that this distribution is also reflective of the rest of the waveform, the estimated average number of random peaks in the 3-μ s integration window is approximately given by 0.75. The Cherenkov cut should therefore provide an excellent rejection of events containing Cherenkov spikes as the probability for additional, spurious SPE within the integration window is low. In order to ensure that the cut is not overly aggressive, which would also remove possible CEνNS events, the performance of the cut was tested with an additional Monte Carlo (MC) study. First, the average width of each peak was calculated using information gathered while determining the mean SPE charge Q_{spe}. The left panel of Fig. 7.9 shows the width distribution of each peak found in each waveform in black. The red fit function is given by:

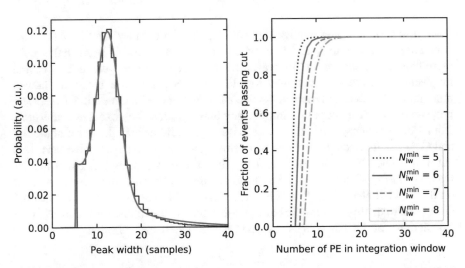

Fig. 7.9 **Left**: SPE peak-width distribution as measured during the barium calibration. **Right**: MC simulation of the Cherenkov cut survival rate for radiation-induced events. A fixed peak width of 13 samples and a fixed SPE charge for each peak are assumed. Including a proper distribution for both parameters leads to a broader rise of the survival fraction but does not diminish the 100% survival rate for larger N_{pe}

$$f\left(q, a_n, \sigma_n, a, \mu, \sigma, k, q_0\right) = \left[a_n e^{-q/\sigma_n} + a\, g(q, \mu, \sigma)\right] \left(1.0 + e^{-k(q-q_0)}\right)^{-1}$$

$$\tag{7.4}$$

$$\text{where}\quad g(q, \mu, \sigma) = \mathrm{Exp}\left[\frac{(q-\mu)^2}{2\sigma^2}\right].$$

which consists of an exponential representing the noise and a single Gaussian representing the width of an individual SPE. The last term reflects the efficiency in detecting a peak with the peak-finding algorithm. The average SPE width is $\Delta t_{spe} = 12.7\,\mathrm{S}$. A $3\,\mu\mathrm{s}{=}1500$ sample long array is populated with N_{pe} PE, where a decay profile following Fig. 5.6 is assumed. Each of these SPE is set to have a fixed width of $13\,\mathrm{S}$. The total number of separate peaks is recorded, and it was checked if current waveform still passed a cut of N_{iw}^{min}. The procedure was repeated 10,000 times for varying the number of photoelectrons N_{pe} and different choices of N_{iw}^{min} and the percentage of simulated waveforms passing the cut was scored. The results are presented in the right panel of Fig. 7.9. The survival rate of the Cherenkov cut quickly rises towards unity and stays at that level for the full cut window, i.e., $N_{pe} \leq 40$. While the Cherenkov cut could be extended to apply to events with higher energy, given the charges to be expected from a Cherenkov spike, the $0 \leq N_{pe} \leq 40$ cut range is deemed sufficient.

7.4.2   Rise-Time Cuts

The second type of data cuts are based on several possible rise-times of the integrated charge of an event. Conceptually, these cuts use the known decay profile of a low-energy radiation-induced event to separate bona fide events from spurious collections of SPE by using rise-time characteristics. Similar PSD approaches are discussed in [3, 4]. Given a scintillation decay profile as shown in Fig. 5.6, the cumulative distribution function of the charge profile can be written as:

$$F_Q(t) = \frac{1}{1+r} F(t, \tau_{fast}) + \frac{r}{1+r} F(t, \tau_{slow}) \tag{7.5}$$

$$\text{with}\quad F(t, \tau) = \frac{1 - e^{-t/\tau}}{1 - e^{-t_{max}/\tau}}.$$

Here, $\tau_{fast} = 527\,\mathrm{ns}$, $\tau_{slow} = 5.6\,\mu\mathrm{s}$, and $r = 0.41$ as shown in [1]. The integration window used in this ^{133}Ba calibration limits $t_{max} = 3\,\mu\mathrm{s}$. As already introduced in Chap. 6, two distinct rise-times are of interest, i.e., T_{0-50} and T_{10-90}. The theoretical threshold crossing times, i.e., T_{10}, T_{50}, and T_{90}, can be calculated by evaluating $F_Q(T_{10}) = 0.1$, $F_Q(T_{50}) = 0.5$, and $F_Q(T_{90}) = 0.9$. The resulting rise-times based on Eq. (7.5) are thus given by:

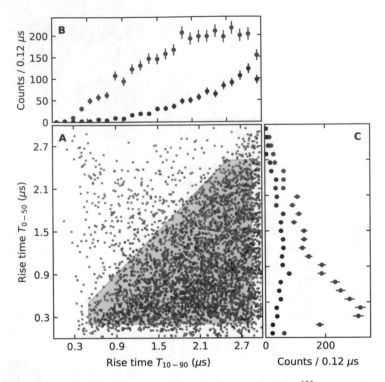

Fig. 7.10 (a) Two-dimensional rise-time distributions for events in the ^{133}Ba calibration data set that pass all quality cuts with an additional Cherenkov cut of $N_{iw}^{min} = 5$. Only events with an energy of $5 \leq N_{pe} \leq 20$ are shown. Red (black) data points represent the $\mathcal{C}$ ($\mathcal{AC}$) data set. The shaded blue region represents one of the proposed rise-time cuts. (**b**) T_{10-90} distribution marginalized over T_{0-50}. (**c**) T_{0-50} marginalized over T_{10-90}. An excess of coincidence events over anti-coincidences is readily visible in all panels

$$T_{0-50} = T_{50} = 506.4\,\text{ns}$$
$$T_{10-90} = T_{90} - T_{10} = 1876.3\,\text{ns}, \tag{7.6}$$

whereas for a flat, i.e., completely random, distribution the expectation is $T_{0-50} = 1.5\,\mu s$ and $T_{10-90} = 2.4\,\mu s$. Therefore, the rise-times provide an excellent method to reject events consisting of spurious, random SPE.

An example of the different rise-time distributions as measured for this ^{133}Ba calibration data set is shown in Fig. 7.10. Panel **a** shows the two-dimensional distribution of rise-times for events with an energy of $5 \leq N_{pe} \leq 20$, that pass all quality cuts and an additional Cherenkov cut of $N_{iw}^{min} = 5$. Red (black) data points represent the $\mathcal{C}$ ($\mathcal{AC}$) data set. Panel **b** shows the T_{10-90} distribution marginalized over T_{0-50} and panel **c** shows T_{0-50} marginalized over T_{10-90}. An excess of coincidence events over anti-coincidences is readily visible in all panels. Panel **b** suggests that the anti-coincidences ($\mathcal{AC}$, black) show a much longer T_{10-90}

rise-time than the events present in the coincidence region (C, red), whereas panel **c** shows that the corresponding $\mathcal{AC}$ T_{0-50} is centered around 1.5 µs. Both of these observations are consistent with the assumption that these anti-coincidence events mainly arise from spurious events.

The shaded blue region in panel **a** represents one of the proposed rise-time cuts, i.e., all events within the polygon are accepted, whereas outside events are rejected. The shape of the area incorporates two main ideas. First, several events above the diagonal of $T_{0-50} = T_{10-90}$ can be seen. As discussed in Chap. 6, the onset of these events was misidentified due to a preceding SPE in the ROI. As a result, the integration window is misplaced and only covers part of the event, which distorts both the absolute charge and the individual rise-times. This cut is termed the diagonal rise-time cut, i.e., $T_{0-50} < T_{10-90}$. Second, additional rise-time cuts are defined that reject fringe cases, i.e., events showing rise-times close to 0 or 3 µs. These cuts form a rectangular acceptance window in the plane spanned by T_{0-50} and T_{10-90}. Therefore, these cuts are referred to as orthogonal rise-time cuts.

To illustrate the effect of the Cherenkov cut on both C and $\mathcal{AC}$ data sets, the rise-time distributions were plotted for different N_{iw}^{min} values. Figure 7.11 shows a rise-time distribution plot similar to the one in Fig. 7.10, where with an increase in

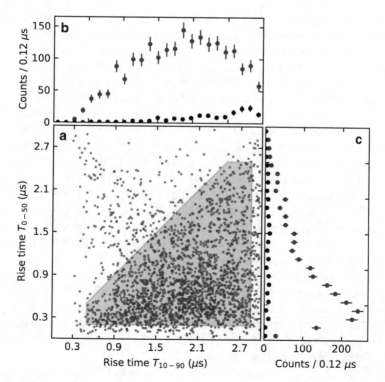

Fig. 7.11 Labeling as in Fig. 7.10. Data quality cuts are identical, whereas the Cherenkov cut was set to $N_{iw}^{min} = 7$. Most of the $\mathcal{AC}$ data is cut, whereas the excess in C remains almost untouched

N_{iw}^{min} from 5 to 7. It is apparent that the $\mathcal{AC}$ data were almost completely rejected by the data quality and the Cherenkov cuts alone. The low-energy events in the $\mathcal{C}$ data set were almost perfectly preserved.

7.4.3   Calculating Cut Acceptances

The signal acceptances for different combinations of the Cherenkov and rise-time cuts need to be quantified. As the procedure is identical for each cut parameter combination, this section describes the acceptance calculation using one particular set of cuts as an example, it is

$$
\begin{array}{lll}
\text{Cherenkov} & N_{iw}^{min} = 8 \\
\text{Orthogonal rise-time cut} & T_{0-50} \in [0.2, 2.5]\mu s \\
\text{Orthogonal rise-time cut} & T_{10-90} \in [0.5, 2.85]\mu s & (7.7) \\
\text{Diagonal rise-time cut} & T_{0-50} < T_{10-90}.
\end{array}
$$

The signal acceptances calculated in the following reflect the acceptances corresponding to the orthogonal rise-time and the Cherenkov cuts and were calculated based on the ^{133}Ba data set. All remaining signal acceptances later used in the CEνNS search (Chap. 9) were calculated using the CEνNS search data recorded at the SNS itself, rather than the ^{133}Ba data set. The rational for this is as follows: First, since the CsI[Na] crystal was exposed to a γ source for the ^{133}Ba calibration, the phosphorescence in the crystal was much higher compared to the CEνNS search data set. Second, the onset of a forward-scattered Compton event is not perfectly aligned with the beginning of the region of interest (ROI) window in the ^{133}Ba calibration measurement. The combination of these two effects creates a larger amount of radiation-induced events in the ^{133}Ba measurement, which are preceded by spurious SPEs from afterglow. The onset for these events is therefore misidentified. As events with a misidentified onset are the main types of events removed by the diagonal rise-time cut, the ^{133}Ba calibration can therefore not be used to predict the percentage of events cut in the CEνNS data set. The diagonal rise-time cut is further discussed at a later point in this thesis.

In order to calculate the signal acceptance for the chosen cut parameters, an energy spectrum calculated from the $\mathcal{AC}$ and $\mathcal{C}$ data sets with only the quality cuts and a diagonal rise-time cut applied was compared to a spectrum with all data cuts applied. The former is termed *uncut* and the latter is termed *cut* spectrum, making reference to the Cherenkov and orthogonal rise-time cuts only. The full list of data cuts applied to each set is shown in Table 7.3.

The resulting $\mathcal{AC}$ and $\mathcal{C}$ energy spectra are shown in the top panel of Fig. 7.12, where an afterglow cut of $N_{pt}^{max} = 8$ was chosen. The *uncut* data is shown in desaturated colors, whereas the *cut* spectra are shown in saturated colors. Several

Table 7.3 Definition of *uncut* and *cut* data sets during the barium calibration

Cut type	Uncut data set	Cut data set
Overflow	✓	✓
Linear gate	✓	✓
Afterglow	✓	✓
Diagonal rise-time	✓	✓
Cherenkov	–	✓
Orthogonal rise-time	–	✓

The naming convention only makes reference to the Cherenkov and orthogonal rise-time cuts

interesting features stand out. First, there is a large excess above 15 PE in the $\mathcal{C}$ region over the $\mathcal{AC}$ region for both *uncut* and *cut* sets. Second, no significant difference between the $\mathcal{C}$ region of both *uncut* and *cut* sets above 15 PE is visible. Third, the rise for both $\mathcal{C}$ and $\mathcal{AC}$ in the *uncut* spectrum at low energies, i.e., below 15 PE, is apparent. In contrast, the *cut* spectra both decline towards zero for $N_{pe} \to 0$ due to the additional Cherenkov and rise-time cuts. Fourth, these additional cuts have a much larger impact on the $\mathcal{AC}$ data than the $\mathcal{C}$ data, confirming that radiation-induced events are mostly preserved, whereas events containing spurious SPEs are rejected.

Once the individual $\mathcal{C}$ and $\mathcal{AC}$ energy spectra had been determined, the residual spectra were calculated as $\mathcal{R} = \mathcal{C} - \mathcal{AC}$ for both *uncut* and *cut* sets. The residuals are shown in the middle panel of Fig. 7.12. As discussed earlier, the residual spectrum only contains small angle Compton-scattered events, as all random coincidences from environmental radiation were subtracted out. The *cut* and *uncut* residual spectra can be compared to find the percentage of events that survive the additional data cuts. The ratio between the *cut* and *uncut* residual spectrum is referred to as signal acceptance fraction, which is shown in the bottom panel of Fig. 7.12.

As noted previously, the average background level of the ^{133}Ba calibration was much higher than of the CEνNS data set due to the presence of the ^{133}Ba source. To confirm that this did not introduce any bias in the signal acceptances derived from this calibration, the acceptance calculation was repeated for all $N_{pt}^{max} \in [3, 10]$. A comparison of the resulting signal acceptance fractions found that no bias was introduced by the choice of N_{pt}^{max}, i.e., the calculated acceptance fractions were independent of N_{pt}^{max}. However, in order to incorporate the uncertainty based on the choice of N_{pt}^{max}, a sigmoid-shaped signal acceptance function $\eta_{ba}(N_{pe})$ was fit to all acceptance fractions (bottom panel of Fig. 7.12) calculated for all $N_{pt}^{max} \in [3, 10]$ simultaneously. The fit model is given by:

$$\eta_{ba}(N_{pe}, a, k, x_0) = \frac{a}{1 + e^{-k(N_{pe}-x_0)}} \Theta_H (N_{pe} - 5), \qquad (7.8)$$

where a describes the maximum amplitude, k the width, and x_0 the location of the sigmoid. Θ_H denotes the Heaviside step function and represents an ultimate lower bound, below which the acceptance is always zero.

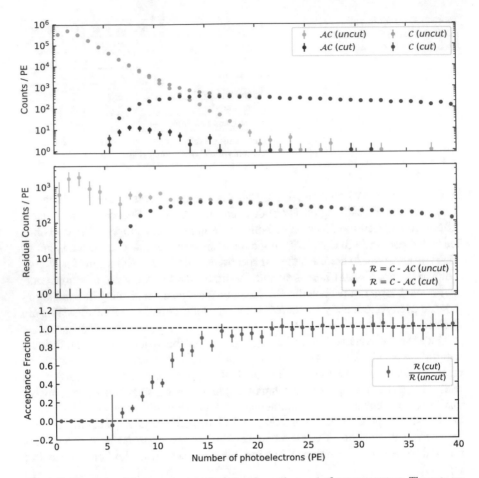

Fig. 7.12 Example of the acceptance calculation for a given set of cut parameters. The cuts are as stated in Eq. (7.7), as well as $N_{\mathrm{pt}} = 8$. **Top**: Energy spectra for both $\mathcal{C}$ and $\mathcal{AC}$, for the *uncut* and *cut* data sets. The *uncut* set is shown in desaturated colors. The roll-off at low energies, i.e., $N_{\mathrm{pe}} \rightarrow 0$ in the *cut* spectra is due to the additional Cherenkov and orthogonal rise-time cuts. **Middle**: Residual spectra $\mathcal{R}=\mathcal{C}-\mathcal{AC}$ for both data sets. The *uncut* residual is shown in desaturated colors. The residual spectrum of each set only includes trigger associated with coincident events as random coincidences were removed by the subtraction of $\mathcal{AC}$. **Bottom**: Acceptance fraction calculated using the ratio between the *cut* and *uncut* residual spectra

The acceptance fractions, i.e., the ratio between the *cut* and the *uncut* residual energy spectrum, are calculated for all $N_{\mathrm{pt}}^{\mathrm{max}} \in [3, 10]$. The resulting acceptances are shown in gray in Fig. 7.13. The best fit of Eq. (7.8) to all of these acceptances simultaneously is shown in solid red.

In order to estimate the uncertainty on the acceptance function, the underlying distributions of a, k, and x_0 were estimated using a bootstrap resampling procedure [5, 6]. First, the model $\eta(N_{\mathrm{pe}})$ was fitted to the data using a least squares approach,

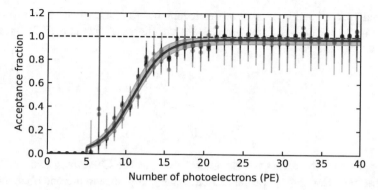

Fig. 7.13 The gray data points represent acceptance fractions calculated from the ^{133}Ba calibration data as shown in Fig. 7.12 for all $N_{pt}^{max} \in [3, 10]$. The red curve represents the best fit of the signal acceptance model $\eta(N_{pe}, a, k, x_0)$ to all data simultaneously. The red band represents the 1σ confidence interval derived using a percentile bootstrapping approach

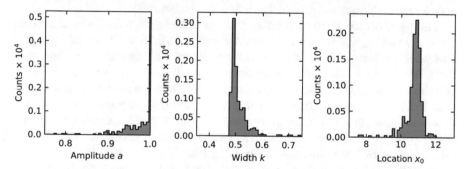

Fig. 7.14 Bootstrapped parameter distributions of a, k, and x_0, i.e., the parameters used in the acceptance function η_{ba} (Eq. (7.8)). This acceptance function quantifies the acceptance of the Cherenkov and orthogonal rise-time cuts using the ^{133}Ba calibration data. The parameter distributions were obtained by using a bootstrap approach of resampling residuals which is further described in the text

and the residuals $\vec{r}$ between the best fit and each acceptance value (gray in Fig. 7.13) were calculated. For each individual acceptance value, a residual was randomly drawn from $\vec{r}$ and added to its original value. As a result, a new synthetic set of acceptance fractions was created. The model $\eta(N_{pe})$ was then refit to the new synthetic set of acceptance fractions, and the resulting fit parameters a, k, and x_0 were recorded. The last two steps, i.e., the creation of a new synthetic set of acceptance fractions and the fit of the acceptance model, were repeated $N_B = 10,000$ times to properly sample the distribution of a, k, and x_0. The result is shown in Fig. 7.14.

These bootstrapped parameter distributions can be used to calculate the 1σ confidence interval of each individual parameter a, k, and x_0, by determining the 15.865 and the 84.135 percentile of the respective bootstrapped distribution [2].

The final 1σ confidence interval of the acceptance function η_{ba} is shown in shaded red in Fig. 7.13. The best fit parameters for the choice of cuts (Eq. (7.7)) presented in this chapter are given by:

$$a = 0.979^{+0.021}_{-0.038}$$

$$k = 0.494^{+0.034}_{-0.013} \tag{7.9}$$

$$x_0 = 10.85^{+0.18}_{-0.40}$$

This approach only covers the acceptances due to the Cherenkov and orthogonal rise-time cuts. Comparing the calculated signal acceptance to the one obtained with a simple MC approach (shown in Fig. 7.9) shows that the real acceptance rises much slower towards unity. However, the acceptance is finite for events with $N_{pe} < 8$. This shows the superiority of the ^{133}Ba measurement over the MC approach in quantifying the signal acceptance function. The following section details how the acceptance fraction of the diagonal rise-time can be obtained.

The signal acceptance was calculated as described above, but with an *uncut* data set where the diagonal rise-time cut was not applied. In this case, the resulting signal acceptance reflects all cuts, Cherenkov, orthogonal, and diagonal rise-time cuts. First, the $\mathcal{R} = \mathcal{C} - \mathcal{AC}$ residuals for both *uncut* and *cut* data sets were calculated, and the resulting energy spectra were compared. Due to the missing diagonal rise-time cut, there are more events in the *uncut* spectrum that are not present in the *cut* spectrum. This results in an overall reduction of the signal acceptance (Fig. 7.15, gray points). As described above, the acceptances calculated for different afterglow cuts N_{pt}^{max} were fitted simultaneously using Eq. (7.8). The best fit is shown as solid red line in Fig. 7.15. The 1σ errors for each individual parameter ($a^\star$, $k^\star$, and $x_0^\star$) were determined with a percentile bootstrap resampling approach as described

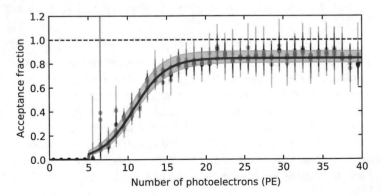

Fig. 7.15 Fit of the signal acceptance model $\eta(N_{pe}, a, k, x_0)$ to acceptances, which were calculated based on an *uncut* residual energy spectrum that did not include a diagonal rise-time cut. As a result, the overall acceptance is smaller than what can be seen in Fig. 7.13

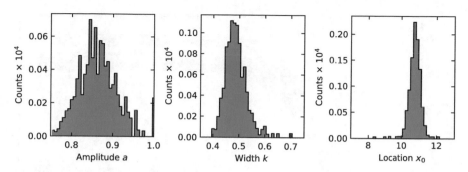

Fig. 7.16 Bootstrapped parameter distributions of $a^\star$, $k^\star$, and $x_0^\star$, i.e., the parameters used in the acceptance function η_{ba} (Eq. (7.8)). This acceptance function quantifies the acceptance of the Cherenkov and orthogonal rise-time cuts using the ^{133}Ba calibration data. The parameter distributions were obtained by using a bootstrap approach of resampling residuals which is further described in the text. The acceptance fractions fitted in this exercise were determined using an *uncut* spectrum that did not include the diagonal rise-time cut

above. The bootstrapped parameter distributions are shown in Fig. 7.16. The fit values for an *uncut* data set without a diagonal rise-time cut is given by:

$$a^\star = 0.845^{+0.059}_{-0.036}$$

$$k^\star = 0.487^{+0.032}_{-0.042} \tag{7.10}$$

$$x_0^\star = 10.76^{+0.21}_{-0.34}$$

It can be observed that both $k^\star$ and $x_0^\star$ remain almost unchanged with respect to the previous approach for which $k = 0.494^{+0.034}_{-0.013}$ and $x_0 = 10.85^{+0.18}_{-0.40}$ were found. However, $a^\star$ is much smaller than $a = 0.979^{+0.021}_{-0.038}$.

In order to scale a to match $a^\star$, the acceptance of the diagonal rise-time cut was calculated separately. Since the percentage of events with a misidentified onset is independent of the energy deposited, high-energy depositions in the CsI[Na] crystal can be used to estimate the overall acceptance η_{drt} of the diagonal rise-time cut. To this end, the fraction of events with $T_{0-50} > T_{10-90}$ in the energy region of 50–150 PE was computed. The same fraction was computed for all $N_{pt} \in [3, 10]$, and a constant was fitted to the data. The best fit is given by:

$$\eta_{drt} = 0.872 \pm 0.011 \tag{7.11}$$

The diagonal rise-time acceptance fraction η_{drt} was then incorporated into the previous acceptance model. It is

$$\hat{\eta}(N_{pe}, a, k, x_0) = \eta(N_{pe}, a, k, x_0) \times \eta_{drt}. \tag{7.12}$$

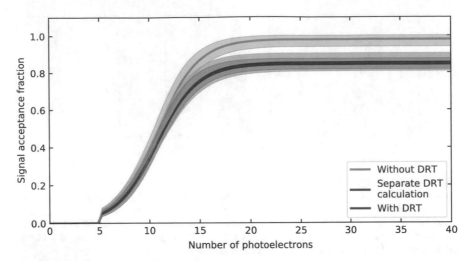

Fig. 7.17 Comparison of different approaches to incorporate the diagonal rise-time cut into the signal acceptance function. Shown in blue is the acceptance model fit using the *uncut* data set that excludes all events with $T_{0-50} \geq T_{10-90}$. The corresponding fit values are given in Eq. (7.9). Black shows the signal acceptance fit obtained for an *uncut* data set that did not exclude events with $T_{0-50} \geq T_{10-90}$. The corresponding fit values are given in Eq. (7.10). The red line shows the blue model after a separate calculation of the diagonal rise-time cut acceptance was included. Once this additional scaling factor was included, the difference between the black and red model is negligible

Figure 7.17 shows a comparison between all signal acceptance models. The blue curve shows the signal acceptance as derived from an *uncut* data set from which all events with $T_{0-50} \geq T_{10-90}$ were removed. The black curve shows the model for an *uncut* data set that includes these events. The red curve represents the blue model after scaling with η_{drt}. As the diagonal rise-time cut only provides an overall scaling of the acceptance function, both amplitudes can be compared. It is

$$a^{\star} = 0.845^{+0.059}_{-0.036} \tag{7.13}$$

$$a \times \eta_{\mathrm{drt}} = 0.979^{+0.021}_{-0.038} \times 0.872 \pm 0.011 = 0.854^{+0.021}_{-0.035} \tag{7.14}$$

The agreement between both approaches is apparent. As a result, the scaling approach in which the signal acceptance η_{drt} of the diagonal rise-time cut is calculated independently was later used to incorporate the acceptance from the diagonal rise-time cut in the CEνNS search data.

References

1. J. Collar, N. Fields, M. Hai et al., Coherent neutrino-nucleus scattering detection with a CsI[Na] scintillator at the SNS spallation source. Nucl. Instrum. Methods Phys. Res. Sect. A Accel. Spectrom. Detect. Assoc. Equip. **773**, 56–65 (2015). http://www.sciencedirect.com/science/article/pii/S0168900214013254
2. P.M. Dixon, Bootstrap resampling, in *Encyclopedia of Environmetrics* (Wiley, Chichester, 2002)
3. X. Luo, V. Modamio, J. Nyberg et al., Test of digital neutron–gamma discrimination with four different photomultiplier tubes for the NEutron Detector Array (NEDA). Nucl. Instrum. Methods Phys. Res. Sect. A Accel. Spectrom. Detect. Assoc. Equip. **767**, 83–91 (2014)
4. E. Ronchi, P.-A. Söderström, J. Nyberg et al., An artificial neural network based neutron–gamma discrimination and pile-up rejection framework for the BC 501 liquid scintillation detector. Nucl. Instrum. Methods Phys. Res. Sect. A Accel. Spectrom. Detect. Assoc. Equip. **610**(2), 534–539 (2009)
5. R. Wehrens, H. Putter, L.M. Buydens, The bootstrap: a tutorial. Chemom. Intell. Lab. Syst. **54**(1), 35–52 (2000). http://www.sciencedirect.com/science/article/pii/S0169743900001027
6. C.-F.J. Wu, Jackknife, bootstrap and other resampling methods in regression analysis. Ann. Stat. **14**, 1261–1295 (1986)

Chapter 8
Measurement of the Low-Energy Quenching Factor in CsI[Na]

As discussed in Chap. 2, the only visible signal from a CEνNS interaction is a nuclear recoil within the detector. These recoils carry only a small amount of energy. Their detection is made even more difficult due to a process typically referred to as quenching: For a low-energy nuclear recoil only a small amount of energy is converted into scintillation or ionization, and the rest is dissipated via secondary nuclear recoils and heat. The quenching factor can be defined by comparing the scintillation or ionization yield of a nuclear recoil of given energy to the scintillation or ionization yield of an ionizing particle of the same energy, i.e., a particle which predominantly loses its energy through electronic recoils in the detector. The quenching factor plays a crucial role in establishing an energy scale for nuclear recoils. It is needed to convert the predicted CEνNS nuclear recoil energies from keV$_{nr}$ to keV$_{ee}$. Using the light yield calibration measured in Chap. 6 using an ^{241}Am source, the electron equivalent energy can further be converted into an equivalent number of photoelectrons N_{pe}, i.e., a quantifiable detector response.

To add to previous measurements of the quenching factor of CsI[Na] two new and independent measurements of the quenching factor were performed in the framework of the COHERENT collaboration at Triangle Universities Nuclear Laboratory (TUNL). The data acquisition system used in the measurement of the CsI[Na] quenching factor described in this chapter was different from the one used in the light yield (Chap. 6) and ^{133}Ba calibration (Chap. 7).

8.1 Experimental Setup

The CsI[Na] quenching factor was measured at Triangle Universities Nuclear Laboratory (TUNL). The experimental setup was located in the shielded source area (SSA) of the facility (Fig. 8.1). Deuterium ions were accelerated by a FN tandem Van de Graaff accelerator and directed towards a deuterium gas chamber. The

© Springer Nature Switzerland AG 2018   81
B. Scholz, *First Observation of Coherent Elastic Neutrino-Nucleus Scattering*,
Springer Theses, https://doi.org/10.1007/978-3-319-99747-6_8

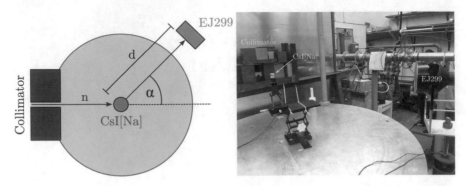

Fig. 8.1 Experimental setup of the CsI[Na] quenching factor measurements at TUNL. A highly collimated neutron beam (red) is directed towards a small CsI[Na] detector (orange). By knowing the angle α between the incoming neutron beam and the EJ299-33A (blue), the energy deposited in the CsI[Na] by a triggering neutron can be kinematically inferred

subsequent reaction of deuterium ions with the deuterated target produced a highly collimated neutron beam of known energy. A small CsI[Na] detector ($l = 55$ mm, $\varnothing = 22.4$ mm), which was procured from the same manufacturer as the CEνNS detector, using an identical growth method and sodium dopant concentration of 0.114 mole %, was positioned in the center of the neutron beam. The size of the detector was chosen such that predominantly single nuclear recoils are produced within the crystal. The CsI[Na] crystal was read out by an ultra-bialkali PMT [12], which made it possible to probe recoil energies down to ~ 3 keV$_{nr}$. The high voltage for the ultra-bialkali PMT was provided by a Stanford Research Systems Inc. PS350 power supply and set to -935 V.

A EJ299-33A plastic scintillator ($l = 47.6$ mm, $\varnothing = 114.3$ mm) was used to detect neutrons that scattered off the CsI[Na] detector. This scintillator was read out by a 5-in 9390B PMT from ET Enterprises. The EJ299-33A is capable of n-γ discrimination using standard pulse shape discrimination (PSD) techniques [16]. EJ299-33A signals produced by nuclear recoils exhibit different scintillation decay times from those produced by electronic recoils. As a result, the percentage of the total scintillation light emitted in the first few nanoseconds of an event is different for nuclear and electronic recoils. This can be used for PSD and is further examined in Sect. 8.2.2. The n-γ discrimination was used in the quenching factor runs to reduce the background caused by triggers on environmental γ radiation. The high voltage for the 9390B PMT was provided by an Agilent E3631A and was set to -750 V. The EJ299-33A output was fed into an Ortec 934 constant fraction discriminator (CFD). The 934 CFD logical output provided the trigger for the U1071A Acqiris 8-bit fast digitizer. The raw output of the ultra-bialkali PMT reading out the CsI[Na] crystal was fed into channel 1 of this fast digitizer, whereas the raw output of the 9390B PMT reading out the EJ299-33A was fed through a 6 dB attenuator and recorded as channel 2. The full energy range of neutron induced events is contained in the digitizer range using the attenuator in the data acquisition system.

For each trigger 6 μs-long waveforms were recorded for both channels, i.e., the raw output of the CsI[Na] and the EJ299-33A detectors. The sampling rate was set to 500 MS s^{-1} with a trigger position set to 2 μs and a digitizer range of ±25 mV. The overall triggering rate for this setup in the environmental radiation field, which consists of mostly γs, was $\mathcal{O}$(250 Hz). Using a ^{22}Na source that emits back-to-back 511 keV annihilation-radiation γ-rays, and which was positioned in the midpoint of the distance between both detectors, an offset of 20 ns between the individual detector channels was found. This offset was corrected for in the analysis.

The simple geometry of this quenching factor measurement allowed for the selection of energy of recoiling cesium and iodine nuclei by varying the scattering angle α. The recoil energy can be calculated using the simple kinematic relation [20]

$$E_R = 2 (1 + A)^{-2} \left(1 + A - \cos^2 \alpha - \cos \alpha \sqrt{A^2 - 1 + \cos^2 \alpha}\right) E_n, \qquad (8.1)$$

where E_n is the energy of the incident neutron, α the scattering angle, and A the mass ratio between the target nucleus and the incoming neutron. Data was taken for a total of seven different runs with different angles α between the incident neutron beam and the EJ299-33A. The quenching factor measurement described in this chapter covered the full energy range of nuclear recoils that is of interest to the CEνNS search at the SNS (Fig. 8.15). The CsI[Na] detector position remained unchanged for every run. The exact EJ299-33A detector locations as well as additional information regarding the expected recoil energy and individual run times are given in Table 8.1. Besides decreasing α, the distance d between both detectors was increased in order to avoid triggering on beam-related neutrons that were not scattered off the CsI[Na]. The following sections will first discuss the calibrations performed for each detector followed by the evaluation of the quenching factor of CsI[Na].

Table 8.1 All EJ299-33A detector positions used in the measurement of the quenching factor of CsI[Na]

Angle α (°)	Distance d (cm)	Recoil energy (keV$_{nr}$)	Run time (min)
45	80	18.44	165
39	90	14.04	182
33	90	10.17	225
27	100	6.86	207
24	100	5.45	178
21	100	4.19	203
18	110	3.09	71

The corresponding recoil energies were calculated using Eq. (8.1). By increasing the distance for smaller angles the triggering of on beam-related neutrons that did not scatter off the CsI[Na] was avoided. The measurement for 18° was cut short due to time constraints imposed by the scheduling of other experiments at TUNL

8.2   Detector Calibrations

8.2.1   CsI[Na] Calibrations

To calibrate the light yield of the CsI[Na] detector a dedicated data set using an
^{241}Am source was taken. For this measurement waveforms were acquired using the
CsI[Na] detector signal as a trigger, rather than the 934 CFD output. The trigger
position was again set to $2\,\mu s$ into the digitized traces. For each trigger the DC
baseline of the corresponding waveform was determined as the median of its first
$2\,\mu s$. The exact pulse onset t^0_{csi} was defined by a threshold crossing of 0.6 mV, i.e.,
three digitizer steps, followed by at least ten consecutive samples above threshold.
The CsI[Na] signal was integrated starting at two samples before t^0_{csi} for a total of
$3\,\mu s$ while ignoring samples below 0.6 mV. The resulting charge spectrum is shown
at the top of Fig. 8.2. The main γ-emission peak at 59.54 keV is readily visible. Due
to the low-energy emission of this isotope and the small detector size most of the
energy depositions occur in close proximity to the source, i.e., close to the surface.
Therefore the K- and L-shell escape peaks can also be resolved. These peaks consist
of two distinct energies, one from Cs and one from I, which are merged due to
the limited energy resolution. In order to be able to test the two competing SPE
charge distribution models described in what follows, the output was at this point
not converted to SPE.

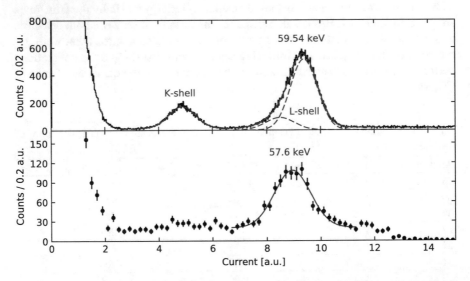

Fig. 8.2 Energy calibration of the CsI[Na] detector used in the quenching factor measurements
at TUNL. **Top:** Dedicated ^{241}Am spectrum taken after the quenching factor measurements. The
^{241}Am main emission peak at 59.54 keV as well as the K- and L-shell escape peaks from Cs and I
can be observed. **Bottom:** Energy calibration using the in-situ ^{127}I$(n, n'\gamma)$ reaction. The drop after
~ 12 a.u. is due to overflows that were excluded from the analysis

Table 8.2 Best fit results for the energy calibration of the small CsI[Na] detector used in the quenching factor measurements at TUNL

Energy (keV)	Centroid μ (a.u.)	Sigma σ (a.u.)
30.93 and 28.57	4.877 ± 0.006	0.498 ± 0.007
55.60 and 55.26	8.495 ± 0.181	0.534 ± 0.063
57.60	8.946 ± 0.033	0.667 ± 0.038
59.54	9.382 ± 0.029	0.499 ± 0.011

The K- and L-shell escape peaks consist of two distinct lines which cannot be resolved due to the limited energy resolution

In addition to the external ^{241}Am calibration, a light yield calibration was performed using the in-situ ^{127}I$(n, n'\gamma)$ reaction. During de-excitation a γ-ray with an energy of 57.6 keV is produced. This light yield calibration uses the data acquired during actual quenching factor runs. As a result the triggering condition was different from the ^{241}Am light yield calibration described above. The DC baseline was determined using the first 1 μs. Only events with a signal onset between 1 μs and 3 μs into the waveform are recorded. This guaranteed that each signal could be integrated over the full 3 μs integration window following the onset. The corresponding charge spectrum is shown at the bottom of Fig. 8.2. Fitting a Gaussian to the peak gave a total of 834 events under the ^{127}I$(n, n'\gamma)$ peak. The best fit results for all peaks are given in Table 8.2.

The calibration results were converted into the light yield of the detector, i.e., the number of PEs produced per keV$_{ee}$ for a given energy. For this purpose the mean SPE charge Q_{spe} for this particular detector-PMT assembly was determined. For each trigger all potential SPE peaks in the pretrace, i.e., the first 1 μs, were identified. An SPE was defined as at least four consecutive samples with an amplitude of at least 0.6 mV. The charge of each SPE was determined by integrating over a corresponding integration window which is defined by two samples before and after the threshold crossings in the sample.

The resulting charge spectrum for the ^{241}Am run can be seen in Fig. 8.3. As discussed in Chap. 6, the mean SPE charge can be extracted by fitting a model to the spectrum. In Chap. 6 the Polya distribution provided a better charge model for the R877-100 PMT used in the CEvNS search. However, the CsI[Na] detector used in the quenching factor measurement described in this chapter consists of a different crystal and used an ultra-bialkali PMT instead of the R877-100 PMT. To determine if the Polya distribution represents a better SPE charge model than a simple Gaussian, a comparative fit procedure is used.

The competing charge models described by Eq. (6.4) (Gauss) and Eq. (6.5) (Polya) were fitted to the SPE charge spectrum. The results are shown in the top panels of Fig. 8.3. The deviation of the model from the data is shown in the bottom panels. The overall deviation of both models is well within 5% for most parts of the fit, except for the low and high charge regions. The discrepancy in the low charge region could indicate under-amplified SPE. The deviation in the high charge region is due to the fact that only contributions from up to two SPE are included in the fit. The mean SPE charge for all other runs was determined in an analog fashion, which all showed the same fit quality.

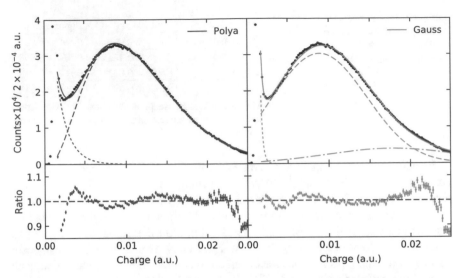

Fig. 8.3 SPE charge spectrum for the ^{241}Am calibration run. Two different models for the charge distribution were fitted to the data. **Left**: The Polya distribution model fails to properly fit the valley between the SPE peak and the rise due to noise. The fit also suggests a negligible contribution from the two SPE peak, which is unphysical given the presence of the ^{241}Am source. **Right**: The Gaussian distribution model performs better at fitting the valley and also includes a significant contribution from the two SPE peak

The variance in the calculated mean SPE charges over all quenching factor runs was negligible. As a result, a single mean SPE charge value is adopted, which is used for all quenching factor runs. It is

$$Q_{\text{spe}}(\text{polya}) = 0.0118 \pm 0.003 \text{ a.u.} \qquad Q_{\text{spe}}(\text{gauss}) = 0.0089 \pm 0.003 \text{ a.u.} \tag{8.2}$$

The energy calibration can be expressed in terms of the light yield, i.e., PEs per keV$_{\text{ee}}$, shown in Table 8.3. The slight increase in light yield around 30 keV compared to that at $\approx$ 60 keV is in agreement with the non-proportional γ-response of CsI[Na] that was measured in [1, 11]. For the remainder of the analysis the light yield obtained from the ^{241}Am line at 59.54 keV was used. This choice was made as the CEνNS detector was also calibrated using a ^{241}Am source (Chap. 6). The nominal light yield at 59.54 keV was linearly extrapolated to very low energies. This avoided any dependence on a particular light emission model and any uncertainty associated. Any energy deposition by nuclear recoils can still be converted directly into units of electron equivalent for any CsI[Na] detector using the quenching factor data, as long as the light yield at $\sim$60 keV is known for a 3 μs long integration window. Previous measurements by Park et al. [13] and Guo et al. [8] used the same approach, which greatly simplifies the direct comparison of the results with their data.

Table 8.3 Light yield calibration for the CsI[Na] detector used in the quenching factor measurements at TUNL

	Polya		Gauss	
Energy (keV$_{ee}$)	Centroid (PE)	Light yield (PE/keV$_{ee}$)	Centroid (PE)	Light yield (PE/keV$_{ee}$)
29.75	413 ± 11	13.88 ± 0.37	548 ± 18	18.42 ± 0.61
55.43	720 ± 24	12.99 ± 0.43	954 ± 38	17.21 ± 0.69
57.60	758 ± 19	13.16 ± 0.33	1005 ± 34	17.45 ± 0.59
59.54	795 ± 20	13.35 ± 0.34	1054 ± 36	17.70 ± 0.60

The energies for the Cs and I K- and L-shell escape peaks were combined to a single central energy assuming an equal contribution from both isotopes

8.2.2 EJ299-33A Calibrations

The EJ299-33A plastic scintillator was calibrated using several different gamma (^{22}Na, ^{137}Cs) and neutron (^{252}Cf) sources. During the calibrations the trigger was set to the EJ299-33A output and the CsI[Na] channel was disregarded. The trigger position was set to $2\,\mu s$ into the $6\,\mu s$ long waveforms. For each trigger the pulse onset was determined by looking for at least ten consecutive samples above the $0.6\,mV$ threshold. For each peak two charge integrals were computed as follows:

$$Q_{\text{long}} = \sum_{t=0\,\text{ns}}^{420\,\text{ns}} V(t) \quad \text{and} \quad Q_{\text{tail}} = \sum_{t=60\text{ns}}^{420\text{ns}} V(t) \tag{8.3}$$

where $V(t)$ represents the digitized voltage values of the EJ299-33A output and $t = 0\,$ns denotes the pulse onset (Fig. 8.6). The long integral Q_{long} is a measure of the total energy deposited in the detector. In contrast, the relative amount of charge contained in the tail integral Q_{tail} depends on the type of the incident particle and is used for n-γ PSD. The timings for both integrals were chosen following [16] to maximize the n-γ discrimination capability for this particular scintillator.

First an energy calibration was performed using the ^{22}Na and ^{137}Cs gamma sources. These provide γ-lines at $511\,$keV and $661.7\,$keV, respectively. Due to the low Z number of both hydrogen and carbon and the relatively low γ energies for both sources, only the Compton edge and not a full energy peak is visible in the spectrum. The top right panel of Fig. 8.4 shows the individual cross-sections for the photoelectric effect, incoherent scattering, and pair production in hydrogen and carbon. It can be observed that the scattering cross-section in hydrogen is almost seven orders of magnitude larger than the one for the photoelectric effect at the energies of interest. In addition, the γ energies are still well below the energy required for pair production. The maximum energy transfer in a single scatter event is given by

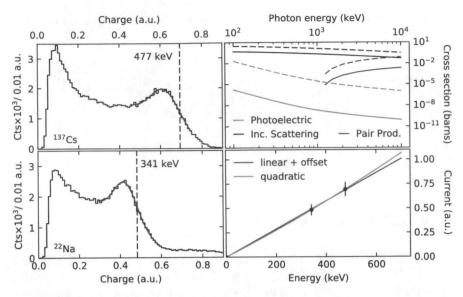

Fig. 8.4 Left top and bottom: Energy calibration spectra recorded with the EJ299-33A detector for two different γ sources. Due to the limited energy resolution the Compton edge is not perfectly well defined. The edge was defined as the 1σ deviation from the peak (dashed lines). **Top right**: Hydrogen (solid) and carbon (dashed) γ-ray cross-sections. Data taken from [2]. **Bottom right**: Linear and quadratic fit to the EJ299-33A calibration data. There is only a minor ($\leq 5\%$) difference between both calibration types for energies below $500\,\mathrm{keV_{ee}}$

$$E_T = E_\gamma \left(1 - \frac{1}{1 + \frac{2E_\gamma}{m_e c^2}} \right). \tag{8.4}$$

The expected Compton edge energies are given by $341\,\mathrm{keV}$ and $477\,\mathrm{keV}$ for a ^{22}Na and ^{137}Cs source, respectively. Due to the poor energy resolution of EJ299-33A, both Compton edges appear as broad peaks. The Compton edge was therefore defined as the 1σ positive deviation from the maximum of the edge (vertical dashed lines in the left, top, and bottom panels of Fig. 8.4). A linear fit to the data points, shown in the right of Fig. 8.4, provides a charge-energy conversion of

$$E[\mathrm{keV}] = 693.95\, Q_{\mathrm{long}}[\mathrm{a.u.}] + 1.13. \tag{8.5}$$

Fitting a quadratic charge-energy relation to the data showed no significant deviation, i.e., $\leq 5\%$, between $30\,\mathrm{keV_{ee}}$ and $500\,\mathrm{keV_{ee}}$. This backs up the linearity of the EJ299-33A detector and substantiates the choices made in the definition of the Compton edges.

During the quenching factor measurements the EJ299-33A triggered with a rate of approximately $250\,\mathrm{Hz}$. Most of these triggers originated from the ambient radiation field, i.e., γs not related to the beam. The environmental background can be reduced by exploiting the n-γ PSD capabilities of the EJ299-33A plastic scintillator. Due to the different decay times [16, 18] for nuclear (13, 50, 460 ns) and electronic (13, 35, 270 ns) recoils, the percentage of charge visible in the tail region of an event is different for these different types of interactions. Following [16], the PSD metric below was adopted:

$$f_{psd} = \frac{Q_{long} - Q_{tail}}{Q_{long}}, \tag{8.6}$$

where Q_{long} and Q_{tail} are defined as in Eq. (8.3). f_{psd} ranges from 0 to 1. Given the aforementioned decay times, events produced by electronic recoils experience a higher f_{psd}.

Two PSD calibration measurements were obtained using a ^{137}Cs and ^{252}Cf source. The ^{137}Cs source only emits γ-rays. As a result all interactions measured in the EJ299-33A in the presence of this source only consist of electronic recoils. When the PSD feature f_{psd} of each event is plotted as a function of its energy E measured in the EJ299-33A these events form a band (black data points in Fig. 8.5). In contrast, the ^{252}Cf source emits both γ-rays and neutrons. As a result both electronic and nuclear recoils are induced in the EJ299-33A. Plotting f_{psd} as a function of the measured energy for this data set results in the formation of two bands. One associated with nuclear recoils and one with electronic recoils (red data in Fig. 8.5). Due to the magnitude of the environmental γ background, a PSD cut was implemented to provide $\geq 99\%$ γ rejection. The cut is shown in shaded blue in Fig. 8.5. The acceptance fraction is shown in the bottom panel of the same figure.

8.3 Quenching Factor Data Analysis

Quenching factor data was taken using the Ortec 934 CFD output as trigger signal for the data acquisition system used in the quenching factor measurements. The trigger position was set to 2 μs into the digitized waveforms (Fig. 8.6). First, a baseline for each trigger was determined for both channels as the median of the first 1 μs. Second, the onset t_{ej} of the EJ299-33A pulse was identified as described in Sect. 8.2. The CsI[Na] trace was scanned for SPE between $t_{ej} - 1$ μs and $t_{ej} + 1$ μs, where an SPE is defined as at least three consecutive samples above 0.6 mV. The first SPE marks the onset t_{csi} of a potential CsI[Na] signal. The lag time is given by $\Delta t = t_{csi} - t_{ej}$, where a negative lag time corresponds to an event with a CsI[Na] onset prior to the trigger. Given a neutron beam energy of ~ 3.8 MeV (Sect. 8.3.2), a lag time of $\Delta t \approx -37$ ns for a measured distance of 1 m between the detectors is expected. After determining t_{csi} the CsI[Na] signal is integrated for a total of 3 μs, where only samples associated with an SPE contribute to the integral. The total charge Q_{total} is converted into a number of photoelectrons N_{pe} using either Q_{spe}(polya) or Q_{spe}(gauss) (Eq. 8.2).

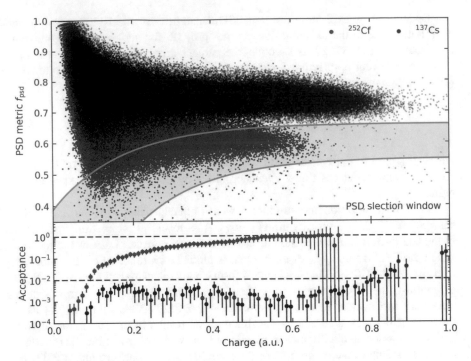

Fig. 8.5 Top: PSD metric f_{psd} as defined in Eq. (8.6) for the EJ299-33A pulse shape discrimination (PSD) calibrations using ^{252}Cf (red) and ^{137}Cs (black). The ^{137}Cs only emits γs and as a result only electronic recoils were induced in the EJ299-33A in the presence of this source. The electronic recoil band covered by the ^{137}Cs data is apparent. The ^{252}Cf emits both γs and neutrons. As a result, events measured in the EJ299-33A in the presence of the ^{252}Cf source are composed of nuclear and electronic recoils. The electronic recoil component is masked by the black data. The nuclear recoil component forms a separate band below the electronic band. The ^{252}Cf data was taken without the 6 dB attenuator. To directly compare it to the ^{137}Cs data set, its signal was divided by 1.995 prior to the analysis. The ^{252}Cf data only covers charges of up to 0.7 a.u. as events with higher charges experienced digitizer range overflows. The shaded blue region shows the PSD cut contour. Only events within this band were accepted as nuclear recoils. **Bottom**: Fraction of events accepted by the PSD cut for both data sets. For a wide range of energies over 99% of the γ-rays from the ^{137}Cs calibration are rejected

8.3.1 Determining the Experimental Residual Spectrum

In order to optimize the signal-to-background ratio, a number of different cuts were applied to the data. First, all events with digitizer range overflow in either channel are excluded. Second, triggers showing more than one peak in the pretrace, i.e., in the first 1 μs, are removed. Third, only events with a neutron-like PSD feature f_{psd} are accepted. Last, a threshold cut on the energy deposited in the plastic scintillator is applied. All of these cuts are independent of the energy deposited in the CsI[Na] and therefore do not contribute to any acceptance bias. The impact of all of these cuts on the data set is exemplified in Fig. 8.7 for the 45° data set. No clear excess is

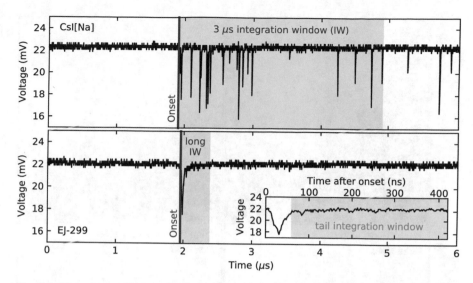

Fig. 8.6 Example waveform taken from the 45° data set of the quenching factor measurement. The CsI[Na] signal is shown on the top, the EJ299 signal on the bottom. The onset (solid red line) in the CsI[Na] crystal happened approximately 34 ns prior to the EJ299-33A onset, compatible with the lag time expected for an event triggered by a scattered neutron. The 3 μs long integration window applied to the CsI[Na] signal is shown in shaded red as is the long integration window Q_{long} for the EJ299-33A. The inset in the lower plot shows the EJ299-33A signal for the full long integration window and highlights the tail integration window Q_{tail} in shaded blue. It can be observed that the main emission is omitted in the tail integral. Due to the different decay times for nuclear and electronic recoils, the ratio between the Q_{long} and Q_{tail} can be used for particle identification as shown in Eq. (8.6) [16]

visible in the top left panel, whereas a clear signal arises after all cuts were applied in the lower right panel at $\Delta t \approx 0\,\mu\text{s}$ and $N_{\text{pe}} \approx 20$.

To define a signal region in Δt, a relevant time limit for which an event in the CsI[Na] could have actually been caused by a scattered neutron needs to be defined. The time-of-flight (ToF) between the CsI[Na] and EJ299-33A detectors can be calculated for non-relativistic neutrons using the energy–velocity relationship

$$E = \frac{1}{2}mv^2 \quad \rightarrow \quad v\left(\frac{\text{cm}}{\text{s}}\right) = 1.3822 \cdot 10^6 \sqrt{E\ (\text{eV})} \qquad (8.7)$$

For a neutron energy of 3.8 MeV (Sect. 8.3.2) and a measured distance between the detectors of one meter, the ToF is approximately 37 ns. However, the spread in neutron energy and the change in distance for different EJ299-33A angles α, the signal window onset was defined as $t_0 = -62\,\text{ns}$ with respect to the EJ299-33A hardware trigger. Due to the low amount of energy deposited in the crystal and the low number of SPE created, the stochastic nature of the light emission of these PE can lead to a nonnegligible spread in the arrival time T_{arr} of the first SPE after the

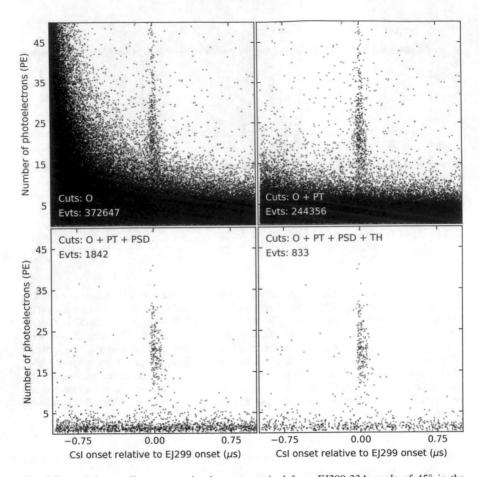

Fig. 8.7 Applying quality cuts on the data set acquired for a EJ299-33A angle of 45° in the quenching factor measurements. The Q_{spe}(gauss) model (Sect. 8.2) was used for the charge to SPE conversion. Negative values on the x-axis indicate events for which the onset in the CsI[Na] occurred before the EJ299-33A trigger. The cuts applied to each data set are shown in the respective plots as is the number of total events passing them. After all cuts are applied, a clear excess around $0\,\mu$s and 20 PE becomes evident. The cuts are abbreviated as follows. **O**: Events showing no digitizer overflow. **PT**: Events with at most one peak in the pretrace. **PSD**: Events showing a neutron-like energy deposition in the EJ299-33A. **TH**: Events with a minimum energy deposition of $400\,\mathrm{keV_{ee}}$ in the EJ299-33A

interaction. Assuming a double exponential scintillation decay profile as measured in [5], the arrival probability density function of an SPE can be written as

$$P_{spe}(t) = \frac{1}{1+r} \frac{\mathrm{Exp}\left[-\frac{-t}{\tau_{fast}}\right]}{\tau_{fast}} + \frac{r}{1+r} \frac{\mathrm{Exp}\left[-\frac{-t}{\tau_{slow}}\right]}{\tau_{slow}}, \tag{8.8}$$

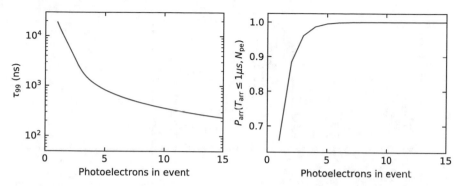

Fig. 8.8 Theoretical arrival times for events in the quenching factor measurements. **Left**: 99% of all events in the quenching factor measurements with a given number of photoelectrons show an arrival time of equal or less than τ_{99}. A rapid decay for $N_{pe} \leq 5$ can be observed. In the CEνNS search (Chap. 9) only events with at least eight individual peaks are accepted. This substantiates that the onset of events in the data is well defined and as a result does not suffer from the spread experienced in this quenching factor measurement. **Right**: Percentage of events with an arrival time of one microsecond or less. Events with at least five N_{pe} are fully contained in the analysis window

where for nuclear recoils $\tau_{fast} = 589 \, ns$, $\tau_{slow} = 6.7 \, \mu s$, and $r = 0.41$ (Fig. 5.6) [5]. The probability of an event with N_{pe} photoelectrons to show an arrival time of $T_{arr} < \tau_i$ can be calculated. It is

$$P_{arr}\left(T_{arr} < \tau_i, N_{pe}\right) = 1 - \left(1 - \int_{t'=0}^{\tau_i} P_{spe}(t')dt'\right)^{N_{pe}}. \tag{8.9}$$

For each $N_{pe} \in [1, 50]$, the time after interaction τ_{99} for which 99% of neutron induced signals already have shown their first PE, i.e., $P_{arr}(T_{arr} < \tau_{99}) = 0.99$, was calculated.

The left panel of Fig. 8.8 shows the evolution of τ_{99} for an increasing number of PE created in an event. A rapid decline in τ_{99} for $N_{pe} \leq 4$ and a more gradual decline for $N_{pe} > 4$ are visible. It also becomes evident that $\tau_{99} > 1.037 \, \mu s$ for $N_{pe} \leq 4$, yet the analysis window is limited to arrival times of $T_{arr} \leq 1.037 \, \mu s$. The right panel of Fig. 8.8 shows the percentage of events exhibiting an arrival time of $1.037 \, \mu s$ or less. For events with 1, 2, 3, or 4 PE this probability, i.e., $P_{arr}(T_{arr} < 1.037 \, \mu s)$, is 65.9, 88.4, 96.0, and 98.6%, respectively. Events with more than four PE exhibit an arrival time of less than $1.037 \, \mu s$ more than 99% of the time. An upper bound for the signal acceptance window can be defined as

$$t_{max}\left(N_{pe}\right) = \begin{cases} t_0 + \tau_{99} + 50 \, ns & \text{for } N_{pe} \geq 5 \\ t_0 + 1.037 \, \mu s + 25 \, ns & \text{otherwise.} \end{cases} \tag{8.10}$$

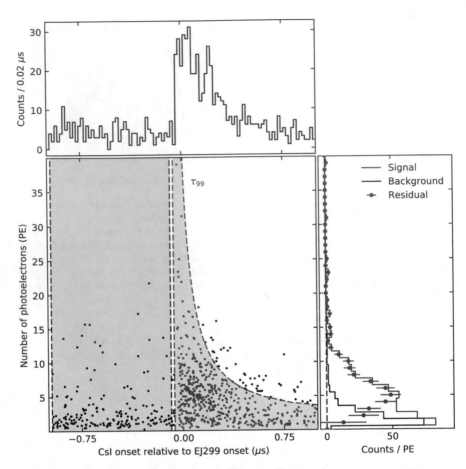

Fig. 8.9 Calculating the residual energy spectrum for the 24° data set of the quenching factor measurement. **Central panel**: Shown are events passing all cuts. The shaded red region marks lag times for possible neutron scatter events. Shaded black regions mark random backgrounds. **Top panel**: The excess caused by neutron recoils at $\Delta t \approx 0$ as well as the spread in arrival times due to the low number of PE involved can be observed. **Right panel**: Energy spectra for both signal (red) and background (black) regions. The background event spectrum was scaled to match the total time over which the signal region was marginalized. The acceptance was corrected for the resulting residual spectrum (blue)

The additional 50 ns in the first case offset the 25 ns buffer introduced in t_0 and provide another 25 ns buffer at the end of the window. However, the signal window for the second case already covers the full analysis window in Δt. 25 ns are added to offset the buffer in t_0. The resulting signal region in Δt is shown in shaded red in Fig. 8.9. By definition at least 99% of events for a given number of PE exhibit arrival times within the signal region. The exception are events with 1–4 PE. In addition, a background region is defined as $\Delta t \in [-975, -80]$ ns, containing random coincidences only, shown in shaded black. Both regions are marginalized over Δt and binned in 1 PE wide bins. Each bin of the resulting background spectrum is scaled to match the exposure of the signal region, i.e.,

$$\eta\left(N_{pe}\right) = \frac{t_{max}\left(N_{pe}\right) - t_0}{895\,ns}. \tag{8.11}$$

The scaled background spectrum is subtracted from the signal spectrum, removing any contribution from random coincidences within the signal region. The resulting residual spectrum therefore only includes events caused by nuclear recoils induced by neutrons scattering off CsI[Na] nuclei.

For events with $N_{pe} \in [1, 4]$ PE the residual needs to be scaled to match the efficiency of all other bins due to the limited analysis window available. A correction factor of $99/65.9$, $99/88.4$, $99/96.0$, and $99/98.6$ was applied for 1–4 N_{pe}, respectively. The right panel of Fig. 8.9 shows an example of all three spectra. The red (black) histogram shows the marginalized spectrum for the signal (background) region, where the background data was already scaled to match the exposure of the signal region. The blue data points show the acceptance corrected residual spectrum.

8.3.2 Simulating the Detector Response Using MCNPX-PoliMi ver.$\tilde{2}$.0

The effective quenching factor for a given recoil energy can be calculated by comparing the experimental residual spectrum with a simulated spectrum of energy depositions within the CsI[Na] detector. Consequently, a comprehensive MCNPX-PoliMi ver. 2.0 [15] simulation spanning 10^9 neutron histories for each of the seven EJ299-33A positions was run. The simulated geometry is shown in the left-hand side of Fig. 8.1, where the collimator is comprised of a concrete wall and a paraffin shield. Both detector geometries were implemented including their specific encapsulation. The energy spectrum of the incident neutron beam was derived by analyzing neutron ToF data taken during the measurements [19], as well as by evaluating simulations of the deuterium–deuterium reaction in the gas cell.

The resulting energy spectrum of the incident neutron beam, which was ultimately implemented in the simulations, is shown in Fig. 8.10. The spectrum peaks at approximately 3.8 MeV and has a FWHM of 0.4 MeV. The two-dimensional beam profile was measured at two different stand-off distances from the gas cell and is shown in Fig. 8.11. A beam profile and spread according to these measurements was included in the simulations.

For each simulated neutron history, scattering at least once in both detectors, the energy deposited at each vertex for both detectors, as well as the type of recoiling nucleus, was recorded. For scatter vertices happening within the EJ299-33A detector, the energy deposited was converted from keV$_{nr}$ to keV$_{ee}$ using a modified Lindhard model for hydrogen recoils [4, 7] and a constant quenching factor of 1% for carbon recoils (Fig. 8.12) [9, 22, 23].

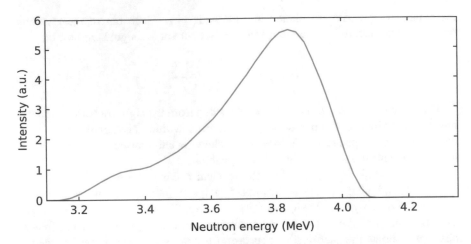

Fig. 8.10 Neutron beam energy spectrum as determined using both ToF measurements and simulations of the deuterium–deuterium reaction in the gas cell. Data provided by Grayson Rich (University of North Carolina at Chapel Hill) [19]. This spectrum was used as the initial neutron energy distribution in the MCNPX-PoliMi ver. 2.0 simulations of the quenching factor measurements

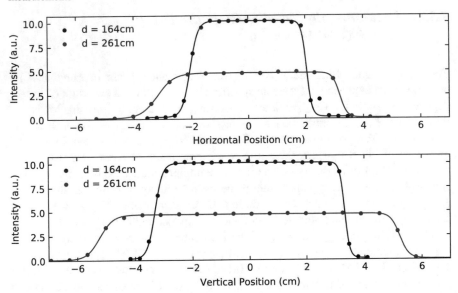

Fig. 8.11 Transversal neutron beam profile available in the shielded source area (SSA) at TUNL as measured at a distance of 164 cm (black) and 261 cm (red) from the tip of the gas cell. The end of the plastic collimator (Fig. 8.1) is located at 145.7 cm. The solid lines represent a fit of two sigmoids to the data in order to estimate the FWHM. The beam divergence according to these measurements was implemented in the MCNPX-PoliMi ver. 2.0 simulations

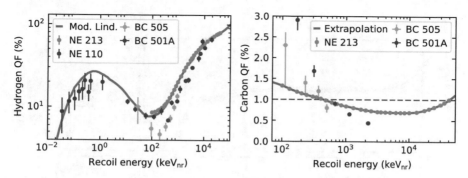

Fig. 8.12 Hydrogen and carbon quenching factors for different organic scintillators. NE-213 is a liquid scintillator and is equivalent to EJ-301 and BC-501 [10]. BC-505 is another liquid scintillator whereas NE-110 is a plastic scintillator [14]. The density of hydrogen and carbon atoms for each of these scintillators is comparable and on the order of 5×10^{22} atom cm^{-3}. Data shown for NE-110 is taken from [7], NE-213 from [22], BC-501A from [23], and BC-505 from [9]. The left panel uses a logarithmic scale, whereas the right panel uses a linear scale. The modified Lindhard model (shown as red curve in the left panel) was adopted for hydrogen recoils. A constant quenching factor of 1% was assumed for carbon recoils

No significant difference was observed for the quenching factor results if the carbon quenching factor is extrapolated as shown in red in Fig. 8.12. This is unsurprising, as the energy transfer is dominated by proton recoils, with approximately 68% of all vertices scatter off of hydrogen and contributing approximately 78% of the total unquenched energy in the EJ299-33A. The sum of all individually quenched recoil energies represents the total energy visible in the plastic scintillator for that particular neutron history.

For the CsI[Na] detector in contrast approximately 93.4% of all histories only scatter once within the crystal due to the small size of the detector (Fig. 8.13 panel **B**). To simplify the analysis a constant quenching factor for each angle is assumed. This allows summing over the energy deposited at each vertex in the CsI[Na] crystal to extract the total energy deposited.

Figure 8.13 shows the results of the simulation corresponding to the 24° data set. Panel **A** shows the visible energy in the EJ299-33A from beam-related neutrons after accounting for the quenching factor of hydrogen and carbon. Panel **D** shows the unquenched energy spectrum deposited in the CsI[Na] crystal. A distinct peak at an energy approximately predicted by Eq. (8.1), 5.45 keV in this case, can be observed. Panel **C** shows the energy deposited in both detectors for each history. The energy deposited in the CsI[Na] is uncorrelated to the energy deposited in the EJ299-33A above $E_{ej299} \gtrsim 50 \, keV_{ee}$. A simple threshold cut is therefore truly energy independent and does not introduce any bias.

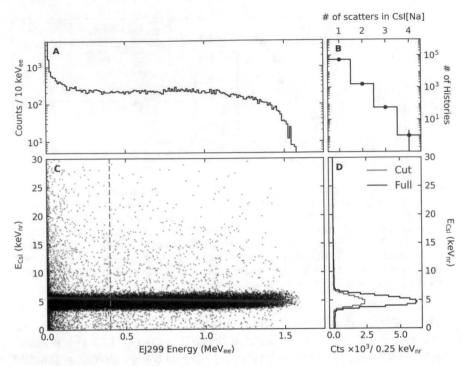

Fig. 8.13 Results of the MCNPX-PoliMi ver. 2.0 simulations for a EJ299-33A position of 24° in the quenching factor measurements. Panel **A** shows the energy depositions within the EJ299-33A detector. The carbon and proton recoils were converted to ionization energies using the known response of a closely related organic scintillator EJ-301 [22] and a modified Lindhard model for low energies [7]. Panel **B** shows the number of times a triggering neutron scatters within the CsI[Na] crystal. Approximately 93% of all neutrons scatter only once. Panel **C** shows the two-dimensional distribution of the unquenched energy deposited in the CsI[Na] detector versus ionization energy deposited in the EJ299-33A. No correlation between the two energies was found. Panel **D** shows the unquenched energy deposited in the CsI[Na] detector. A distinct peak arises from the kinematic selection of scattered neutrons. No change in the peak shape and location is visible if events below 400 keV$_{ee}$ (dashed red line in the bottom left plot) are cut

8.3.3 Extracting the Quenching Factor

To calculate the best estimate for the quenching factor Q_f at a given energy, the simulated recoil spectrum was compared to the experimental residual spectrum. First, the nuclear recoil energy of every simulated interaction in the CsI[Na] was converted into an electron equivalent energy using a constant quenching factor Q_f. The electron equivalent energy was further converted into a number of photoelectrons using the light yield calculated earlier (Sect. 8.2.1). For each neutron history k scattering m_k times within the CsI[Na] detector, the average number μ_{pe}^k of PE produced is given by

$$\mu_{\text{pe}}^k = \sum_{i=1}^{m_k} Q_f \, \mathcal{L} \, E_i^k \tag{8.12}$$

where Q_f represents the quenching factor, $\mathcal{L}$ is the light yield, and E_i^k is the energy deposited at vertex i for the neutron history k. A Poisson spreading was applied to every μ_{pe}^k to calculate the simulated energy spectrum n. For each bin j, where j is the number of PE produced, it is

$$n^j = \sum_k P\left(j|\mu_{\text{pe}}^k\right) \quad \text{with} \quad P\left(x|\mu\right) = \frac{e^{-\mu}\mu^x}{x!}. \tag{8.13}$$

The simulated spectrum needs to be scaled to match the exposure acquired in the experiment, i.e., the number of neutrons incident on the CsI[Na] need to match in both cases. The final simulated energy spectrum N_{sim} is therefore given by

$$n_{\text{sim}}^j = n^j \frac{t_\alpha \phi_n}{10^9}, \tag{8.14}$$

where t_α is the time measured for angle α in seconds and ϕ_n the neutron flux provided at the neutron beam facility. The χ^2 goodness-of-fit statistic for the chosen quenching factor and neutron scaling is

$$\chi^2 = \sum_j \frac{\left(n_{\text{res}}^j - n_{\text{sim}}^j\right)^2}{n_{\text{res}}^j}, \tag{8.15}$$

where n_{res}^j and n_{sim}^j represent the experimental residual and the simulated spectrum, respectively. The best fit is given by the minimum of the two-dimensional χ^2 distribution.

Figure 8.14 shows this process for the 24° data set. The two-dimensional χ^2 distribution is shown on the left. A white dot represents the minimum and 1- and 2-σ contours are shown as solid and dashed white lines, respectively, where the 1- and 2-σ contours are defined by an increase of χ^2 with respect to the minimum by $\Delta\chi^2 = 2.30$ and $\Delta\chi^2 = 4.61$, respectively [17]. The right panel shows the best fit of the simulated recoil spectrum (red) to the experimental residual data (blue). The excellent agreement between simulation and experimental data is readily visible. The fit quality for all other data sets is similar to the one presented.

To properly propagate any uncertainties, the fitting procedure was repeated by varying the SPE mean charge and the light yield within their respective errors. To explore the uncertainty associated with the choice of t_0 and t_{max} the signal window was varied by ±24 ns. Further the PSD cut contours were floated by ±2% and different threshold cuts were applied on the minimum total energy deposited in the EJ299-33A, i.e., $E_{\text{threshold}} \in [250, 300, 350, 400, 450, 500]$. The best fit values for

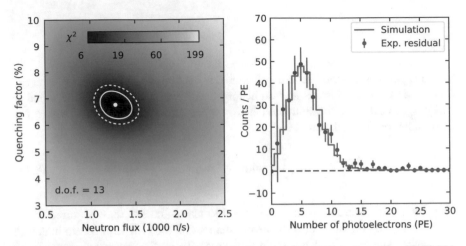

Fig. 8.14 Fitting the simulated recoil data to the experimental residual spectrum for a EJ299-33A position of 24° in the quenching factor measurements. **Left**: Two-dimensional χ^2 distribution for the quenching factor fit. The white dot marks the best fit. The solid and dashed white lines represent the 1- and 2-σ contours, respectively, i.e., $\Delta\chi^2 = 2.30$ and $\Delta\chi^2 = 4.61$ [17]. **Right**: Best fit of the simulation data (red histogram) to the residuals (blue). The residual data was calculated as shown in Fig. 8.9

Table 8.4 CsI[Na] quenching factor and neutron flux for all EJ299-33A positions

Angle (°)	Recoil energy (keV$_{nr}$)	Quenching factor Q (%)	Neutron flux ϕ_n (1000 n/s)
18	2.87 ± 0.56	5.21 ± 1.74	1.20 ± 0.43
21	4.04 ± 0.72	6.38 ± 0.76	1.15 ± 0.33
24	4.80 ± 0.80	6.75 ± 0.75	1.31 ± 0.25
27	6.18 ± 0.90	6.90 ± 0.65	1.34 ± 0.34
33	9.07 ± 1.21	7.38 ± 0.51	1.77 ± 0.39
39	12.61 ± 1.43	7.14 ± 0.67	1.24 ± 0.41
45	16.37 ± 1.81	7.16 ± 0.63	1.50 ± 0.44

The quenching factor data is also shown in Fig. 8.15 where previous measurements by others were included. The neutron flux calculated in this analysis is shown in Fig 8.16, including the flux derived from an in-situ measurement of the inelastic $^{127}I(n, n'\gamma)$ reaction rate

Q_f and ϕ_n as well as the minimum χ^2 were recorded for each of these results. The best estimate of the quenching factor Q_f and the neutron flux ϕ_n is given by the best fit found for the initial choice of cut parameters. In contrast, the uncertainty represents the square root of the unweighted variance of all fit results added in quadrature to the 1-σ uncertainty of the initial best fit. The final best fit values for each angle are given in Table 8.4.

Figure 8.15 shows the measurement presented in this thesis in blue. A decreasing quenching factor for low-energy recoils can be seen, where previous measurements suggested a rise (black, green). The quenching factor measurement provided in [5] was not included as the quenching factor definition employed there slightly differed

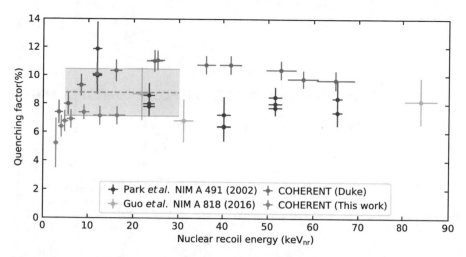

Fig. 8.15 Comparison of the CsI[Na] quenching factor measurement with previous experiments. The results of this thesis are shown in blue. Previous measurements are shown in black [13] and green [8]. A second measurement of the quenching factor performed by the COHERENT collaboration is shown in red. The data for the second measurement was also acquired at TUNL with the same crystal-PMT assembly and neutron source, but differed in data acquisition system and data analysis. The grayed region spans the energy region of interest for the CEνNS search presented in Chap. 9. The dotted line represents the choice of quenching factor used in the CEνNS search. The shaded area shows the corresponding 1σ uncertainty band

from the one used in the experiments shown here. A second measurement performed by the Duke group of the COHERENT collaboration (red) found a slightly higher quenching factor, but also observed a decreasing scintillation efficiency at low energies. As semi-empirical quenching models fail to reproduce this behavior [21], a constant quenching factor is assumed for the energy region of interest in the CEνNS search discussed in this thesis. To define this constant quenching factor, all available data in the CEνNS search region, i.e., $\sim$5-30 keV$_{nr}$, was included in the calculation. The lower bound of this energy range is set by the threshold achieved in the CEνNS search (Chap. 9). The upper bound represents the energy at which the CEνNS recoil rate becomes negligible. The best fit value was determined by weighing each data point with its uncertainty [6]. To estimate the uncertainty on the best fit quenching factor the unweighted standard deviation of all data points was used. The final quenching factor is given by

$$Q_f = 8.78 \pm 1.66\%. \tag{8.16}$$

Figure 8.15 shows the quenching factor adopted in the CEνNS search as a dotted brown line. The uncertainty band is shown as a shaded brown region.

Figure 8.16 shows the best fit neutron flux derived from the scaling factor (Eq. 8.14) for each angle. The large uncertainty of the scaling factor mainly arises from the uncertainty of the PSD cut efficiency. The efficiency shown in Fig. 8.5

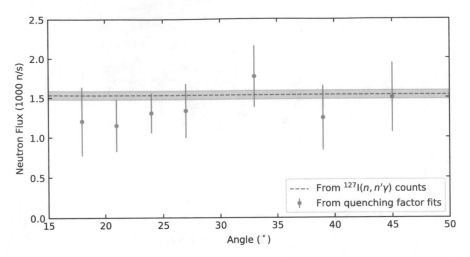

Fig. 8.16 Best fit neutron flux derived for each fit. The large uncertainties are due to the uncertainty in the cut efficiency of the PSD. The average neutron flux derived from the inelastic counts is 1534 ± 53 neutrons per second, compatible with all positions at the 1σ level

does not reflect the true efficiency for accepting a neutron induced event as the ^{252}Cf calibration still contains a large portion of γ-like triggers. As the acceptance simply compares the number of events surviving the cut after marginalizing over f_{psd}, the true efficiency should be larger. This error increases for smaller threshold energies. Therefore, the best fit probably slightly underestimates the neutron flux. The constant neutron flux derived for every angular position of the EJ299-33A detector substantiates that no threshold effects were introduced in this analysis [3].

The true neutron flux variation between runs was estimated to be less than 10% by Grayson Rich, being compatible with the variation shown for most angles, the exception being 33°. Due to the large uncertainties on the neutron flux, this deviation appears not to be concerning. In addition to the neutron flux estimates from the scaling factor of the quenching factor fits, an in-situ measurement of the inelastic scattering rate ^{127}I$(n, n'\gamma)$ can be used to estimate the neutron flux incident on the detector. Due to the triggering condition employed (beginning of Sect. 8.3), the coincidence rate Γ_c of an inelastic scatter event in the CsI[Na] detector being visible in the analysis window, i.e., $t_{ej} - 1\,\mu s$ and $t_{ej} + 1\,\mu s$ (beginning of Sect. 8.3), can be calculated using

$$\Gamma_c = \Gamma_t \Gamma_i \Delta t \tag{8.17}$$

Here $\Gamma_t \approx 240 - 265$ Hz is the average total triggering rate for each angular EJ299-33A position, Γ_i the rate of inelastic scatter events in the CsI[Na], and $\Delta t = 2\,\mu s$ is the width of the analysis window in each waveform. Γ_i was calculated using a dedicated MCNPX-PoliMi ver. 2.0 simulation including 10^9 neutron histories and is dependent on the exact neutron flux ϕ_n delivered by the accelerator. It is

$$\phi_n = \frac{N_c \, N_n}{N_i \, N_t \, \Delta t},$$ (8.18)

where $N_c = 834$ is the number of inelastic scatter events found in the analysis that coincide with the trigger. $N_n = 10^9$ is the total number of neutrons simulated, $N_i = 14,518,617$ the number of inelastic events produced in the simulation, and $N_t = 18,723,150$ the total number of triggers taken in all of the individual quenching factor runs. The average neutron flux throughout all measurements is given by $\phi_n = 1534 \pm 53$ neutrons per second, which is shown in Fig. 8.16 as a red band.

This secondary measurement again suggests that the true neutron flux was slightly underestimated in the quenching factor fits. However, as the neutron flux only provides an overall scaling factor this slight bias is considered negligible.

References

1. P.R. Beck, S.A. Payne, S. Hunter et al., Nonproportionality of scintillator detectors. V. comparing the gamma and electron response. IEEE Trans. Nucl. Sci. **62**(3), 1429–1436 (2015), http://ieeexplore.ieee.org/document/7118262/
2. M. Berger, J. Hubbell, S. Seltzer et al., *XCOM: Photon Cross Section Database (Version 1.5)* (National Institute of Standards and Technology, Gaithersburg, MD, 2010). Online. http://physics.nist.gov/xcom
3. J.I. Collar, A realistic assessment of the sensitivity of XENON10 and XENON100 to light-mass WIMPs. arXiv preprint arXiv:1106.0653 (2011)
4. J.I. Collar, D. McKinsey, Comments on "First Dark Matter Results from the XENON100 Experiment". arXiv preprint arXiv:1005.0838, https://arxiv.org/abs/1005.0838v3 (2010)
5. J.I. Collar, N. Fields, M. Hai et al., Coherent neutrino-nucleus scattering detection with a CsI[Na] scintillator at the SNS spallation source. Nucl. Instrum. Methods Phys. Res. Sect. A Accelerators Spectrom. Detect. Assoc. Equip. **773**, 56–65 (2015), http://www.sciencedirect.com/science/article/pii/S0168900214013254
6. J.I. Collar, D. Akimov, J.B. Albert et al., Observation of coherent elastic neutrino-nucleus scattering. Science **357**(6356), 1123–1126 (2017), http://science.sciencemag.org/content/357/6356/1123
7. D. Ficenec, S. Ahlen, A. Marin et al., Observation of electronic excitation by extremely slow protons with applications to the detection of supermassive charged particles. Phys. Rev. D **36**(1), 311 (1987)
8. C. Guo, X. Ma, Z. Wang et al., Neutron beam tests of CsI(Na) and CaF2(Eu) crystals for dark matter direct search. Nucl. Instrum. Methods Phys. Res. Sect. A Accelerators Spectrom. Detect. Assoc. Equip. **818**, 38–44 (2016), http://www.sciencedirect.com/science/article/pii/S0168900216002047
9. J. Hong, W. Craig, P. Graham et al., The scintillation efficiency of carbon and hydrogen recoils in an organic liquid scintillator for dark matter searches. Astropart. Phys. **16**(3), 333–338 (2002)
10. LIQUID SCINTILLATORS. Online. http://www.eljentechnology.com/products/liquid-scintillators. Accessed 26 June 2017

11. W. Mengesha, T. Taulbee, B. Rooney et al., Light yield nonproportionality of CsI (Tl), CsI (Na), and YAP. IEEE Trans. Nucl. Sci. **45**(3), 456–461 (1998)
12. K. Nakamura, Y. Hamana, Y. Ishigami et al., Latest bialkali photocathode with ultra high sensitivity. Nucl. Instrum. Methods Phys. Res. Sect. A Accelerators Spectrom. Detect. Assoc. Equip. **623**(1), 276–278 (2010)
13. H. Park, D. Choi, J. Choi et al., Neutron beam test of CsI crystal for dark matter search. Nucl. Instrum. Methods Phys. Res. Sect. A Accelerators Spectrom. Detect. Assoc. Equip. **491**(3), 460–469 (2002), http://www.sciencedirect.com/science/article/pii/S0168900202012743
14. PLASTIC SCINTILLATORS. Online. http://www.eljentechnology.com/products/plastic-scintillators. Accessed 26 June 2017
15. S.A. Pozzi, E. Padovani, M. Marseguerra, MCNP-PoliMi: a Monte-Carlo code for correlation measurements. Nucl. Instrum. Methods Phys. Res. Sect. A Accelerators Spectrom. Detect. Assoc. Equip. **513**(3), 550–558 (2003), http://www.sciencedirect.com/science/article/pii/S0168900203023027
16. S. Pozzi, M. Bourne, S. Clarke, Pulse shape discrimination in the plastic scintillator EJ-299-33. Nucl. Instrum. Methods Phys. Res. Sect. A Accelerators Spectrom. Detect. Assoc. Equip. **723**, 19–23 (2013), http://www.sciencedirect.com/science/article/pii/S016890021300524X
17. W.H. Press, S.A. Teukolsky, W.T. Vetterling et al., *Numerical Recipes in C: The Art of Scientific Computing (Cambridge)* (Cambridge University Press, Cambridge, 1992)
18. Pulse shape discrimination EJ-299-33A, EJ-299-34 (2016). Online. http://www.eljentechnology.com/index.php/products/plastic-scintillators/ej-299-33a-ej-299-34
19. G. Rich, Measurement of low-energy nuclear-recoil quenching factors in CsI[Na] and statistical analysis of the first observation of coherent, elastic neutrino-nucleus scattering. Ph.D. thesis (2017)
20. Y. Shimizu, M. Minowa, H. Sekiya et al., Directional scintillation detector for the detection of the wind of WIMPs. Nucl. Instrum. Methods Phys. Res. Sect. A Accelerators Spectrom. Detect. Assoc. Equip. **496**, 347–352 (2003), http://www.sciencedirect.com/science/article/pii/S0168900202016613
21. V. Tretyak, Semi-empirical calculation of quenching factors for ions in scintillators. Astropart. Phys. **33**(1), 40–53 (2010), http://www.sciencedirect.com/science/article/pii/S0927650509001650
22. V. Verbinski, W. Burrus, T. Love et al., Calibration of an organic scintillator for neutron spectrometry. Nucl. Instrum. Methods **65**(1), 8–25 (1986), http://www.sciencedirect.com/science/article/pii/0029554X68900037
23. S. Yoshida, T. Ebihara, T. Yano et al., Light output response of KamLAND liquid scintillator for protons and 12 C nuclei. Nucl. Instrum. Methods Phys. Res. Sect. A Accelerators Spectrom. Detect. Assoc. Equip. **622**(3), 574–582 (2010)

Chapter 9
CEνNS Search at the SNS

The previous chapters discussed all calibration and background measurements performed prior to the CEνNS search at the SNS, which is described in this chapter. The CsI[Na] detector was described in Chap. 5, which also included a schematic of the data acquisition system and the data format. The SNS beam facility was discussed in detail in Sect. 3.1. Between June 25th, 2015 and May 26th, 2017 data was almost continuously acquired at the SNS with a 60 Hz triggering rate. The total number of triggers acquired in this time period was 2,825,705,648.

Two 70 μs long waveforms were acquired for each protons-on-target (POT) trigger. The first channel contains the CsI[Na] signal, the second contains the discriminated output from the muon veto panels (Fig. 5.3). Both traces were sampled at 500 MS s^{-1}, resulting in 35,000 sample long waveforms. The trigger position was set to 78.5% of the total waveform, i.e., at sample 27,475. As discussed in Sect. 5.1, the POT trigger, i.e., *event 39*, provided by the SNS was used as external trigger. The POT trigger occurs at a constant rate of 60 Hz regardless of the operational status of the SNS.

For about two-thirds of the recorded data, the SNS was operational, i.e., protons were impinging on the mercury target and neutrinos were produced. For the remaining third the SNS underwent planned maintenance and no neutrinos were emitted. For the remainder of this thesis, the subset of data for which neutrinos were produced is referred to as $\mathcal{ON}$ data set. Likewise, periods for which no power on target was provided are referred to as $\mathcal{OFF}$ data set.

9.1 Waveform Analysis

A large fraction of the CEνNS search data analysis mimicked the analysis presented in Chap. 7. This enabled the ^{133}Ba calibration (Chap. 7) to be used to quantify the acceptances of Cherenkov and rise-time cuts employed in the CEνNS search. Two

© Springer Nature Switzerland AG 2018
B. Scholz, *First Observation of Coherent Elastic Neutrino-Nucleus Scattering*,
Springer Theses, https://doi.org/10.1007/978-3-319-99747-6_9

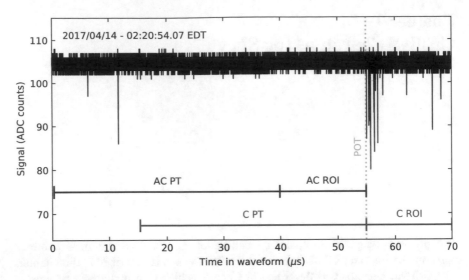

Fig. 9.1 Example CsI[Na] waveform highlighting the different analysis regions used in the CEνNS search data analysis. Also shown is the POT trigger provided by the SNS. The AC ROI cannot contain any beam-related events and was used to estimate the event rate from environmental backgrounds. The long PT regions provided a cut against contamination from afterglow from preceding high-energy events

overlapping sub-regions were defined for each individual CsI[Na] waveform, i.e., the coincidence (C) and anti-coincidence (AC) region. The analysis pipeline used for both C and AC region was identical. As a result, the AC region can be used to estimate the environmental steady-state background present in the C region. The subtraction of C-AC therefore only includes beam-related events (Sect. 9.4.2).

Both C and AC are further subdivided into a PT and ROI. The positioning of AC and C was chosen such that the region of interest (ROI) of the former ends at the protons-on-target (POT) trigger, whereas the ROI of the latter begins there. This choice guaranteed that no potential CEνNS signal could be present in the AC ROI. The AC region can therefore be used to estimate the random, environmental background occurring before the POT trigger. These random backgrounds include Cherenkov light emission in the PMT window, spurious SPEs from afterglow, and random groupings of dark-current photoelectrons, among others. A waveform highlighting the different regions is shown in Fig. 9.1. The exact start and end sample of each region are given in Table 9.1.

For each waveform it was first determined whether it was fully contained within the digitizer range. An overflow ($f_o = 1$) was recorded if at least one sample showed an amplitude of $+127$ or -128 ADC counts in either the CsI[Na] and/or the muon veto channel. Second, a linear-gate flag ($f_g = 1$) was recorded if the number of falling (n_f) and rising (n_r) threshold crossings of the 18 ADC counts level were unequal in the CsI[Na] signal, i.e., $n_f \neq n_r$.

Table 9.1 Start and end
sample of each analysis
region in the CEνNS search

	AC region		C region	
	Start	End	Start	End
PT	150	19,950	7675	27,475
ROI	19,950	27,475	27,475	35,000

The starting sample was included in the
region, whereas the ending sample was
excluded

The global baseline V_{median} of the CsI[Na] signal was estimated using the median of the first 20,000 samples. The CsI[Na] signal V was baseline shifted and inverted using

$$\hat{V}_i = V_{median} - V_i \quad \text{for} \quad i \in [0, 35,000). \tag{9.1}$$

The location of each peak p_i was determined using the same peak detection algorithm described in Chap. 6. A peak is defined as at least four consecutive samples with an amplitude of at least 3 ADC counts. Both positive and negative threshold crossings were recorded for each peak and its charge calculated as described in Chap. 6. All charges are added to an SPE charge distribution on a 10 min basis, i.e., a new distribution was created for each 10 min time interval. Each charge distribution was fitted using Eq. (6.5), providing an independent mean SPE charge Q_{spe} for each time interval.

Once all peaks were integrated and added to the corresponding charge spectrum the individual C and AC regions were analyzed. The analysis for both of these regions was identical and as such no distinction is made in the following description. The analysis of each individual region was almost identical to the one used during the ^{133}Ba calibration measurements (Sect. 7.2). The only difference between the waveform analysis for the ^{133}Ba calibration data acquired at the University of Chicago and the CEνNS search data acquired at the SNS is the exact onset of each analysis window (C and AC) and the length of the PTs. The analysis method is illustrated in Fig. 7.6. The exact analysis windows for the ^{133}Ba calibration are defined in Table 7.1, whereas the analysis windows used for the CEνNS search data are presented in Table 9.1.

In a first step, the respective regions (C or AC) were extracted from the waveform and the PT and ROI sub-regions defined. The total number of peaks present in the PT was recorded. The location of the first peak within the ROI was determined, which defined the onset of the 1500 sample $= 3\,\mu s$ long integration window. As already shown in Sect. 8.3.1, CsI[Na] represents a fast scintillator (Fig. 5.6). As a result, the onset of an event in the CsI[Na] is well defined as long as more than four peaks are produced (Fig. 8.8). The CEνNS search described in this chapter used a Cherenkov cut of $N_{iw}^{min} = 8$, which demands the presence of at least eight peaks in the integration window. This ensured that the uncertainty on the onset is negligible.

A new baseline was determined for the integration window using the median of the $1\,\mu s$ immediately preceding the window. The integration window was extracted from the ROI and shifted using the new baseline. The amplitude of any sample, that did not belong to any peak p_i found by the peak detection algorithm, was set to zero. The scintillation curve $Q(t)$ was calculated integrating over the zero suppressed signal (Eq. 7.2). The total charge Q_{total} was recorded and the T_{10}, T_{50} and T_{90} rise-times were determined as described in Chap. 6 and shown in Fig. 6.3. The parameters extracted from each CsI[Na] waveform are identical to the ones determined in the ^{133}Ba calibration (Chap. 7) and are given in Table 7.2.

In contrast to the data acquired for the ^{133}Ba calibration the CEνNS search data set also included the muon veto channel. For each trigger the location of all muon veto events present in the trace was determined, using the standard peak-finding algorithm described in Chap. 6. The logical signal of the discriminated muon veto output was detected if at least ten samples showed a minimum amplitude of 10 ADC counts above baseline. If a muon is present in either C or AC a muon veto flag ($f_m = 1$) was raised for the trigger. If there was any muon present in the waveform, the position of the first muon event was recorded. A muon was recorded in approximately 1% of all triggers recorded (panel **e** of Fig. 9.2).

All parameters were recorded in a separate and independent n-tuple for the AC and C region. The procedure described above was repeated for each waveform, where all n-tuples calculated based on the AC region were added to the $\mathcal{AC}$ data set and all n-tuples calculated for the C region were added to the $\mathcal{C}$ data set. Even though the analysis program was optimized to extract all necessary parameters in two passes of each waveform, it still required several weeks on a computer cluster to analyze all of the $\sim$ 2.8 billion triggers acquired. The analysis was performed on the HCDATA cluster provided by the Physics Division at ORNL. The analysis program fully made use of the multi-core processing capabilities provided by the cluster. Approximately 20–30 analysis jobs were continuously running in parallel on HCDATA, using the same amount of CPUs. The full analysis code used on HCDATA is available at [13].

9.2 Detector Stability Performance over the Full Data-Taking Period

Once all data was processed, the detector stability over the 2 years of operation was investigated in order to ensure a high data quality for all time periods used in this analysis. Several different metrics were calculated using the n-tuples created for each waveform, all of which are shown in Fig. 9.2. In the following sections each panel is discussed in detail.

9.2.1 Beam Energy Delivered on the Mercury Target

Panel (**a**) of Fig. 9.2 shows the total daily beam energy delivered by the SNS on the mercury target. Several distinct features are apparent. First, three time periods were excluded from the analysis, shown in shaded gray. The first two were excluded, as

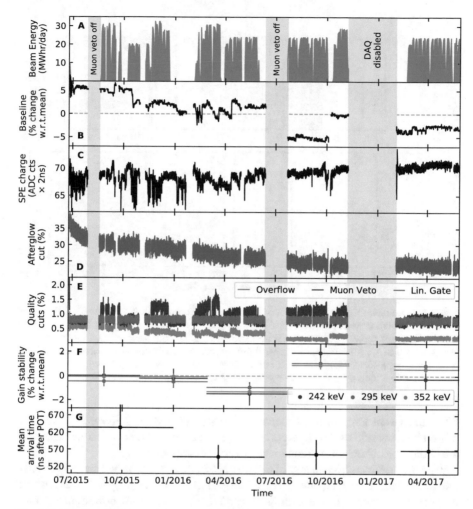

Fig. 9.2 Detector stability tests performed during the 2 years of CsI[Na] data-taking at the SNS. Panel (**a**) is described in Sect. 9.2.1. Panel (**b**) and (**c**) are examined in Sect. 9.2.2. An in-depth discussion of panel (**d**) is provided in Sect. 9.2.3. Panel (**e**) is discussed in detail in Sect. 9.2.4. Lastly, panel (**f**) and (**g**) are examined in Sect. 9.2.5

the muon veto was not operational during these dates. Both of these incidents were linked to power outages at the SNS after which the high voltage supply for the muon veto did not properly restart. The last gap, in contrast, is linked to a failure of the digitizer used in the data acquisition system occurring after the full data acquisition system was restarted. The NI 5153 was sent off to the manufacturer for repairs. Once this issue was resolved the data acquisition was restarted and no further issues were found. Second, gaps in beam energy delivered are apparent, which are not linked to any issues within the experimental setup. These represent times during which the

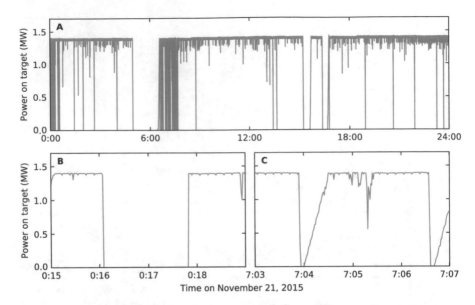

Fig. 9.3 Exemplary beam power retrieved from the SNS archives for November 21, 2015. Panel (**a**) shows the beam power on target for the full day. Short and long drops in beam power are visible. Panel (**b**) shows a section of bad beam power data, i.e., the recorded data instantaneously jumps from 0 to 1.4 MW without a proper ramp-up process. SNS personnel pointed out that this behavior is unphysical. Panel (**c**), in contrast, shows a beam power drop followed by a proper ramp-up process. This drop was therefore deemed physical and included in the analysis

SNS underwent long-term maintenance and no POTs were delivered. Third, dips within beam windows are apparent, which are caused by outages that last less than a full day, e.g., short-term maintenance every Tuesday morning.

The SNS delivered an energy of up to 30 MWh on the SNS target, per day in 2015 and early 2016. However, due to several target failures the beam power was reduced to provide a more stable operation at ∼24 MWh. The beam power information is available from the SNS archive server with a timing resolution of 1 s. A post-processing step was applied to the retrieved beam power information data, which is described in the following.

Panel (**a**) of Fig. 9.3 shows an example beam power time series retrieved from the archive for November 21st, 2015. The recorded power occasionally drops to zero and instantly recovers (panel **b**). Such a behavior is unphysical for an accelerator and to be addressed. For each drop in beam power the corresponding rise in beam power at the end of the gap is analyzed. Whenever an instantaneous rise back to full operational power is visible the gap was flagged as unphysical and excluded from the analysis. In contrast, if the beam was found to be properly ramped back up to full power (panel **c**) the period in question was flagged as physical and a zero power delivered on target was recorded for the gap duration. Less than 0.1% of the triggers acquired in the CEνNS search data set had to be excluded from the analysis.

Once these beam power drops were properly addressed, the CsI[Na] data acquired at the SNS was split into $\mathcal{ON}$ and $\mathcal{OFF}$ data sets on a second by second basis. Using this information the total beam energy delivered on the mercury target was calculated for the whole $\mathcal{ON}$ data set. It is

$$E_{beam} = 7475 \, \text{MWh}. \tag{9.2}$$

This information is used in Sect. 9.4.1 to scale the CEνNS interaction rate prediction in the CsI[Na].

9.2.2 Baseline and Mean SPE Charge Stability

Panel (**b**) of Fig. 9.2 shows the change in the CsI[Na] baseline over the 2 years of data-taking. The data shown is based on 10 min averages of the baseline determined using the first 40 μs of each waveform. The dashed line represents the average for the full 2 years of data-taking.

Panel (**c**) of Fig. 9.2 shows the time evolution of the mean SPE charge in 10 min intervals. Several short-term drops in the SPE charge are apparent. These drops are not correlated with a drop in baseline. Figure 9.4 shows the charge spectra recorded during such a drop. The charge spectra were fitted using Eq. (6.5). Panel (**a**) shows the 10 min interval directly preceding the SPE charge drop. The average SPE is large enough such that less than 1% of SPEs are missed using the standard peak-finding algorithm. In addition these PE only carry a charge less than one-third of Q_{spe}. The bias introduced is therefore negligible.

Panel (**b**) highlights the spectrum calculated during the interval containing the SPE charge drop. The full distribution is shifted towards lower charge values, i.e., the reduced charge is not the result of any fit error. The mean charge dropped by almost 20%. A much higher number of SPEs are missed in the low charge tail.

Panel (**c**) shows the slow recovery after the initial sharp drop and panel (**d**) shows the full recovery of the mean SPE charge. For almost all cases a drop in charge only lasts for a singular 10 min interval, after which the charge recovers as shown. The cause of this behavior was not determined, however, one possibility was a faulty power supply. The original Ortec 556 power supply which provides the high voltage for the CsI[Na] detector was replaced with another unit of the same kind in 2017. No more drops in the mean SPE charge were recorded since then. As these events happened infrequently enough, all periods with a mean SPE charge of < 62 ADC counts $\times$ 2 ns were excluded from the data analysis.

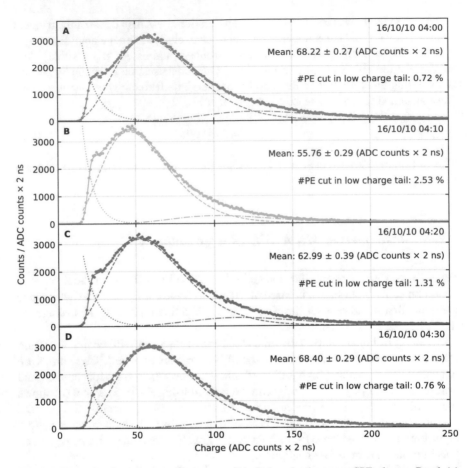

Fig. 9.4 Example charge spectra for an occasional drop in the mean SPE charge. Panel (**a**) shows the charge distribution immediately preceding a recorded drop. Panel (**b**) and (**c**) show the distribution during the drop, whereas panel (**d**) shows the distribution once the charge has fully recovered. These drops are therefore not the result of badly converged SPE charge fits. However, as shown in the text, these drops in SPE charge happened only infrequently and were later attributed to transients in the Ortec 556 power supply providing the high voltage for the R877-100 PMT

9.2.3   Afterglow Cut and the Decay of Cosmogenics

Panel (**d**) of Fig. 9.2 shows the percentage of events rejected by an afterglow cut of $N_{\mathrm{pt}}^{\max} = 3$. The overall behavior is the same for all choices of $N_{\mathrm{pt}}^{\max} \in [1, 10]$. The rejection percentage decays over time. This decline consists of two different time scales, one on a much shorter than the other. Both decay times were determined by fitting the afterglow acceptance time evolution with

$$f(t, a_s, \tau_s, a_l, \tau_l, f_0) = a_s \mathrm{e}^{-t/\tau_s} + a_l \mathrm{e}^{-t/\tau_l} + f_0. \tag{9.3}$$

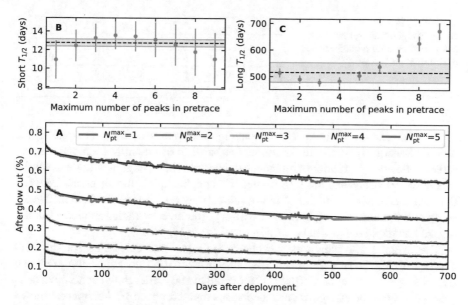

Fig. 9.5 Determining the decay times apparent in the afterglow cut. Panel (**a**) shows the percentage of events rejected by the afterglow cut for different choices of N_{pt}^{max} (color). The rejection percentage decays over time for all choices of N_{pt}^{max}. Shown in black are fits of Eq. (9.3) to the respective data. Panel (**b**) shows the short half-life for all choices of N_{pt}^{max}. The uncertainty weighted average is shown as dashed black line. The 1σ error is shown as a shaded gray band. Panel (**c**) shows the same for the long half-life

Panel (**a**) of Fig. 9.5 shows the time evolution of the afterglow cut for multiple choices of N_{pt}^{max} (colored). The black lines represent the corresponding fits of Eq. (9.3). Panel **b** (**c**) shows the short (long) half-life determined for different choices of N_{pt}^{max}. The dashed black line shows the uncertainty weighted average of the data. The shaded gray region represents the associated 1σ uncertainty. The decay times were found to be

$$T_{1/2}^{short} = 12.77 \pm 0.39 \, \text{days}$$

$$T_{1/2}^{long} = 515.83 \pm 36.92 \, \text{days}$$

As the CsI[Na] detector was driven from the underground laboratory at the University of Chicago to ORNL there was a period of approximately one day over which the detector was above ground and during which cosmogenics could were created. The short decay time is compatible with cosmogenic ^{126}I production as the half-life of this isotope is given by $T_{1/2}(^{126}\text{I}) = 12.93 \pm 0.05$ days [1]. The long decay time is probably dominated by cosmogenic-origin neutron capture on ^{133}Cs, as the half-life of the associated isotope is given by $T_{1/2}(^{134}\text{Cs}) = 2.0648$ years [5, 10].

Table 9.2 Beam energy-weighted afterglow cut acceptances for a number of different choices of N_{pt}^{max}

N_{pt}^{max}	$\eta_{afterglow}$	N_{pt}^{max}	$\eta_{afterglow}$	N_{pt}^{max}	$\eta_{afterglow}$
0	0.167	4	0.817	7	0.910
1	0.405	5	0.863	8	0.923
2	0.606	6	0.891	9	0.932
3	0.738				

This calculation did not use the decaying event rate from a distinct peak in the recorded energy spectrum to determine the half-life of potential cosmogenics, but rather relies on the random coincidences between the afterglow from a potential ^{134}Cs- or ^{126}I-decay and the SNS trigger. This might explain the difference between the measured half-life and the literature value. However, after approximately 2 years of operation under an overburden of 8 m.w.e., the level of radiation present in the crystal has stabilized as can be seen in panel (**d**) of Fig. 9.2.

The average afterglow cut acceptance $\eta_{afterglow}$ was calculated for different choices of N_{pt}^{max}. This acceptance is used in Sect. 9.4.1 to adjust the predicted CEνNS signal rate in the CsI[Na]. The overall average afterglow cut acceptance is calculated from the afterglow cut acceptances found for each 10 min interval shown in Fig. 9.2. Since the CEνNS event rate is directly correlated to the beam energy delivered on the SNS target, the afterglow cut acceptances found for each 10 min interval are weighted by the beam energy delivered in that interval. It is

$$\eta_{afterglow} = \frac{\sum_t E_B(t)\, \eta_{afterglow}(t)}{\sum_t E_B(t)} \qquad (9.4)$$

where $\eta_{afterglow}(t)$ denotes the fraction of events accepted in the 10 min interval t and $E_B(t)$ represents the total integrated power on target delivered in the same window. This only includes time periods within the $\mathcal{ON}$. The exact beam energy averaged acceptances for different N_{pt}^{max} is given in Table 9.2

9.2.4 Quality Cuts

Panel (**e**) of Fig. 9.2 shows the time evolution of events rejected by the different quality cuts. The red line represents the percentage removed due to a linear gate being present in the CsI[Na] signal (Sect. 5.1). The rejection rate remained constant over the full data acquisition period. This agrees with the assumption that most of the high-energy events being cut by the linear gate were induced by cosmic-ray muons traversing the crystal.

The black line represents the percentage of events rejected due to a coincident event in the muon veto waveform. A clear correlation between the energy delivered on the mercury target (panel **a**) and the muon veto rejection rate is visible. This correlation arises from the close proximity of the MOTS exhaust pipe to the muon veto panels, which was described in Sect. 5.3.

Finally, the blue line represents the percentage of events rejected due to digitizer overflows. A correlation between the overflow rate and the average baseline level (panel **b**) is apparent. The closer the baseline is to the upper digitizer range the more triggers are rejected due to overflows.

A beam energy-weighted acceptance fraction is calculated for each quality cut. $\eta_{muon\text{-}veto}$, $\eta_{linear\text{-}gate}$, and $\eta_{overflow}$ were defined following Eq. (9.4). The beam energy-weighted acceptance fractions are given by

$$\eta_{muon\text{-}veto} = 0.989 \qquad \eta_{linear\text{-}gate} = 0.992 \qquad \eta_{overflow} = 0.997 \qquad (9.5)$$

9.2.5 Light Yield and Trigger Position

The data presented in panels (**f**) and (**g**) of Fig. 9.2 were both calculated by Alexey Konovalov at the National Research Nuclear University MEPhI. Panel (**f**) shows the gain stability of the CsI[Na] detector using γ-lines from backgrounds internal of the crystal, namely ^{212}Pb and ^{214}Pb [6]. These γ-rays carry energies beyond the digitizer range. Alexey estimated the total energy of each event exhibiting an overflow by removing the clipped part of the scintillation curve and measuring the charge in the tail only. The gain remained stable within $\sim 1.5\%$ over the course of the 2 year period.

Panel (**g**) shows the stability of the POT trigger provided by the SNS by searching for prompt neutron interactions in the muon veto panels. Even though these panels cover a large area, their neutron efficiency is severely limited due to the high discriminator threshold. Only a small excess of events caused by prompt neutrons is visible in the muon veto over the steady-state background. The low count rate is largely responsible for the large uncertainty on the measurement. The arrival times of the prompt neutrons following the POT trigger derived using this method are similar to those determined in Chap. 4.

A replica of Alexey's analysis is shown in Fig. 9.6, which included a total of 4651 h of $\mathcal{ON}$ data. An excess from prompt neutrons shortly following the POT trigger is apparent. As only the arrival time of the first muon was recorded, Fig. 9.6 shows an excess for low muon arrival times.

9.3 Monitoring Prompt Neutrons Using the Inelastic Scattering ^{127}I(n,n'γ) Reaction

In the previous section the arrival time after the POT trigger was measured for beam-related, prompt neutrons using the muon veto event distribution. The event rate from these prompt neutrons in the CsI[Na] was found to be small in Chap. 4.

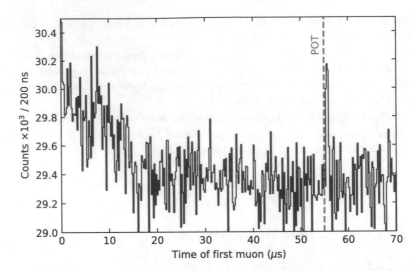

Fig. 9.6 Muon veto hit distribution showing a prompt neutron component shortly following the POT trigger in the CEνNS search data. The excess at the beginning of the trace represents a bias in the analysis as only the onset of the first muon veto hit within each waveform was recorded. However, this bias was deemed unimportant as the purpose of this measurement was to detect beam-related prompt neutrons

However, these calculations heavily depended on neutron transport simulations using MCNPX-PoliMi ver. 2.0. In this section a test of the accuracy of these simulations is provided using the inelastic scattering $^{127}I(n,n'\gamma)$ reaction. Due to its large cross-section [6, 8], this process provides an excellent monitoring tool for prompt neutrons following the POT that were able to penetrate the $\sim 20\,m$ of shielding between the SNS target and the CsI[Na] detector (Fig. 3.3). Using the ^{241}Am source to calibrate the CsI[Na] detector (Chap. 6) ensured that the main gamma emission of the inelastic scattering reaction at 57.6 keV was initially contained within the digitizer range. However, due to a drift in the DC baseline of the CsI[Na] detector signal it is no longer fully contained. In the following sections a calibration measurement using a ^{252}Cf neutron source and a search for the inelastic scattering $^{127}I(n,n'\gamma)$ reaction in the CsI[Na] caused by prompt neutrons associated with the SNS protons-on-target are described.

9.3.1   Validating Neutron Transport Simulations Using a ^{252}Cf Source

The prompt neutron event rate presented in Chap. 4 strongly depended on MCNPX-PoliMi ver. 2.0 simulations. It is therefore crucial to confirm that these simulations are accurate. One way to test the accuracy of these simulations is to use an external neutron source and to compare the simulated prediction to the experimental measurement. As such, a ^{252}Cf neutron source with a yield of $\sim$ 8600 neutrons per second was placed on the outside of CsI[Na] shielding, i.e., outside of the water tanks. The emission spectrum above 1 MeV for this neutron source shows a comparable hardness to the prompt neutron spectrum found in Chap. 4 [14]. Its emitted neutrons carry enough energy to penetrate the neutron moderator and are still able to undergo inelastic scattering within the detector.

However, such a neutron source could potentially activate the detector or shielding material. This calibration measurement was therefore performed at the beginning of a scheduled, month-long SNS maintenance so that any potentially activated materials could decay away before new $\mathcal{ON}$ data was acquired. In addition to this, the full CEνNS search data set described in this thesis was acquired prior to this calibration measurement to avoid any possible contamination.

One of the potential long-lived backgrounds that could be produced in this ^{252}Cf calibration is ^{134}Cs due to neutron capture on ^{133}Cs. As already discussed in Sect. 9.2.3, the half-life of this isotope is approximately 2 years. However, a rough estimation shows that this potential background is negligible: Assuming that all of the neutrons emitted from the ^{252}Cf source actually thermalize within the moderator between the source and the CsI[Na] detector and that no neutron is captured in between a total of $\sim$160 neutrons per second would traverse the CsI[Na] detector. All of the neutrons are assumed to capture as the capture cross-section is $\mathcal{O}(50\,\mathrm{b})$ [11]. For a total calibration period of 2 h this would result in the creation of 1.1×10^6 ^{134}Cs isotopes. Approximating a linear decay of half of these isotopes over a two 2 period results in 0.01 decays per second. Assuming each of these events contaminates a total of 1 ms due to its afterglow a rate of random coincidences between the decaying ^{134}Cs and the POT trigger is expected on the order of

$$\Gamma_c = \Delta t\,\Gamma_{\mathrm{sns}}\,\Gamma_{\mathrm{Cs\text{-}134}} = (70\,\mu\mathrm{s} + 1\,\mathrm{ms}) \times 60\,\mathrm{Hz} \times 0.01\,\mathrm{Hz} = 6.42 \times 10^{-4}\,\mathrm{Hz}, \tag{9.6}$$

which is negligible.

The ^{252}Cf source was placed on the outside of the CsI[Na] shielding, in the middle of the side of water tanks facing the MOTS pipe at a height of 60 cm measured from the floor (Fig. 5.1). The data acquisition was set to trigger directly on the CsI[Na] channel, similar to what is described in Chap. 6. Data was acquired for a total of 107 min. The overall data analysis closely follows the light yield calibration presented in Chap. 6. The measured energy spectrum is shown in the top-panel of Fig. 9.7.

In contrast to the light yield calibration no digitizer overflows were excluded from the data set. This was necessary as a significant part of the events in the inelastic peak actually exceeded the digitizer range. As a result events showing a digitizer overflow had to be included in the data analysis to correctly determine the total number of inelastic scattering events in the CsI[Na]. The energy for events exceeding the digitizer range was corrected using the following approach.

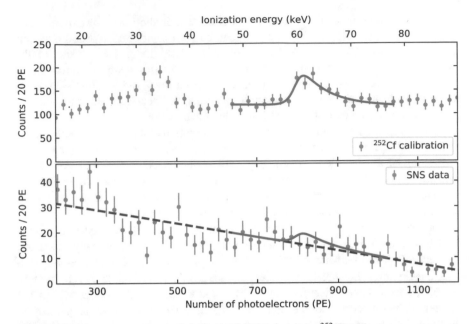

Fig. 9.7 **Top**: Energy spectrum recorded in the CsI[Na] during the ^{252}Cf calibration measurement. The *shark-tooth* shaped peak at approximately 60 keV is due to inelastic neutron scattering ^{127}I(n,n'γ). The unique shape and the shift towards higher energies is caused by the addition of the energy from the recoiling nucleus as well as γ de-excitation [7, 9]. The second peak at approximately 30 keV is caused by the electron capture decay of ^{128}I. The red curve represents the fit of an *ad hoc* peak template (Eq. 9.10) to the data. **Bottom**: Energy spectrum recorded in the CsI[Na] for events with an arrival time of 200–1100 ns, i.e., the arrival time window for prompt neutrons (Chap. 4). Shown are only events taken from $\mathcal{C}$ in the full $\mathcal{ON}$ data set. A linear background model was fitted to the whole spectrum, shown in dashed gray. The fit was used as background in the *ad hoc* peak template, which was fitted to the data. The only free parameter in this fit was the peak amplitude. Shown in red is the 90% confidence limit for this fit

First, events were simulated for different energies, i.e., N_{pe}, using the light yield calibration (Chap. 6) and light emission profile as shown in Eq. (8.8). The light emission profile for CsI[Na] slightly changes with energy. Using the ^{241}Am calibration data the following parameters were found for Eq. (8.8) for an energy of approximately 60 keV$_{ee}$

$$r = 0.691 \qquad \tau_{\text{fast}} = 309\,\text{ns} \qquad \tau_{\text{slow}} = 1649\,\text{ns}. \tag{9.7}$$

The simulated events were clipped at 209 ADC counts to mimic the digitizer range overflow. This particular value represented the total range available given the average baseline value during the ^{252}Cf calibration.

The left panel of Fig. 9.8 shows a simulated light emission profile in black and the corresponding truncated profile in dashed red. Both light curves are integrated and the total number of photoelectrons N_{pe} measured in both cases was compared.

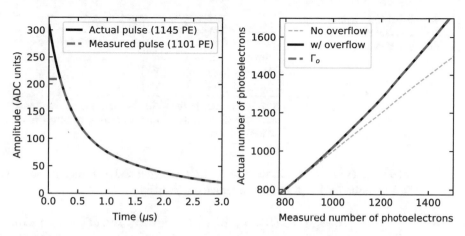

Fig. 9.8 Correcting the energy of events experiencing digitizer overflows. **Left**: The simulated light curve for an event with an energy of 1145 PE is shown in black. The corresponding truncated waveform due to digitizer range overflow is shown in dashed red. Only the first ∼400 ns exceed the digitizer range. The charge cut by the truncation (44 PE = 1145 − 1101 PE) is only a small fraction of the full integrated charge. **Right**: Comparison of the total number of photoelectrons in an event and the number of photoelectrons measured from integrating a truncated event. The deviation between both measures below ∼ 1000 PE is negligible. The red dashed line represents a fit of Eq. (9.8) to the relationship

The evolution of the ratio between true and measured energy is shown in the right panel of Fig. 9.8. The following empirical model Γ_o was fitted to the data

$$\Gamma_o(N_{\text{pe}}, a_1, a_2) = N_{\text{pe}} + \Theta_H(N_{\text{pe}} - 775) \left[a_1 \left(N_{\text{pe}} - 775 \right) + a_2 \left(N_{\text{pe}} - 775 \right)^2 \right].$$
(9.8)

The best fit parameters are given by

$$a_1 = -0.014 \qquad a_2 = 4.5 \times 10^{-4}.$$
(9.9)

The difference between the N_{pe} calculated for the full light emission and the N_{pe} calculated for the truncated emission was found to be negligible up to ∼ 1000 PE. However, the visible energy of all events showing a digitizer range overflow was adjusted using Γ_o.

The resulting energy spectrum for the ^{252}Cf calibration is shown in the top panel of Fig. 9.7. A peak from inelastic scattering ^{127}I(n,n'γ) at around 60 keV is apparent. The small shift towards higher energies as well as the *shark-tooth* shape can be explained by the addition of the energy of the recoiling nucleus and additional γ de-excitation [7, 9]. The second peak at ∼ 30 keV is caused by the electron capture decay of ^{128}I ($E_\gamma \sim 31$ keV) which is produced by thermal neutron capture [5].

Table 9.3 *Ad hoc* template fit parameters for the californium calibration

$I(N_{pe}, a, N_0, N_1, \delta_N)$	$B(N_{pe}, m, c)$
$a = 69.1 \pm 7.1$	$m = (-1.4 \pm 4.6) \times 10^{-2}$
$N_0 = (7.99 \pm 0.10) \times 10^2$	$c = (1.29 \pm 0.34) \times 10^2$
$N_1 = (8.20 \pm 0.38) \times 10^2$	
$\delta_N = 68 \pm 39$	

The fit parameters found for N_0, N_1, and δ_N were used to fix the shape of the ^{127}I(n,n'γ) peak within the CEvNS search data set

The total number of inelastic scattering events was calculated by fitting an *ad hoc* peak template to the peak region. This template is given by

$$\zeta(N_{pe}) = I(N_{pe}, a, N_0, N_1, \delta_N + B(N_{pe}, m, c) \tag{9.10}$$

$$I(N_{pe}, a, N_0, N_1, \delta_N) = a \left[1 + e^{-0.1(N_{pe} - N_0)} \right]^{-1} e^{-(N_{pe} - N_1)/\delta_N} \tag{9.11}$$

$$B(N_{pe}, m, c) = mN_{pe} + c \tag{9.12}$$

where I represents the inelastic peak and B a linear background. The exact fit parameters are given in Table 9.3. The total number of inelastic scattering events was determined by integrate the contribution from I alone, which yielded

$$N_{is}^{exp} = 589 \pm 68 \tag{9.13}$$

The number of inelastic scattering events calculated above was further compared to the prediction based on an MCNPX-PoliMi ver. 2.0 simulation. The number of inelastic scattering events expected was calculated as follows: First, the total number of neutrons emitted from the ^{252}Cf source was calculated. The source was acquired 2 months prior to this measurement and its activity was given as $A = 2\,\mu$Ci $\pm 10\%$ by the manufacturer. The branching ratio of spontaneous fission is given by 3.092% [5] and the number of neutrons per fission is 3.7692 [3]. The half-life of ^{252}Cs is given by $T_{1/2} = 2.645$ year [5]. The total neutron yield Γ_n can therefore be written as

$$\Gamma_n = 2\,\mu\text{Ci} \times 0.03092 \times 3.7692 \times \text{Exp}\left[-\frac{0.167 \text{ year}}{2.645 \text{ year} \ln(2)} \right] = 7873\frac{n}{s} \pm 10\% \tag{9.14}$$

Second, the efficiency of a single neutron emitted by the ^{252}Cf source to actually undergo inelastic scattering within the CsI[Na] detector was calculated using an MCNPX-PoliMi ver. 2.0 simulation. A comprehensive MCNPX-PoliMi ver. 2.0 simulation was performed and this efficiency was found to be $\eta_{is} = 1.31 \times 10^{-5}$. The total number N_{is}^{sim} of inelastic scattering ^{127}I(n,n'γ) events expected based on these simulations is given by

$$N_{is}^{sim} = \Delta T \times \Gamma_n \times \eta_{is} = 6420\,s \times 7873\frac{n}{s} \times 1.31 \times 10^{-5} = 662 \pm 10\%.$$

$$(9.15)$$

The number of experimental counts $N_{is}^{exp} = 589 \pm 68$ is compatible with the predicted number $N_{is}^{sim} = 662 \pm 66$ within their respective errors. The *shark-tooth* shape of the inelastic scattering ^{127}I(n,n'γ) peak was also correctly predicted by the simulations. This exercise confirmed the validity of the neutron transport simulations used in Chap. 4.

9.3.2 Inelastic ^{127}I(n,n' γ) Scattering in the CEνNS Search Data

Using the inelastic scattering ^{127}I(n,n'γ) reaction, bounds on the maximum background rate caused by prompt neutron interactions in the CsI[Na] can be determined in addition to the ones derived in Chap. 4. The high-energy spectrum, i.e., 200–1200 PE, is calculated for the full $\mathcal{C}$ data in the $\mathcal{ON}$ data set. Based on the arrival time spectrum of prompt neutrons measured in Chap. 4, only events with an arrival time of $T_{arr} \in [200, 1100]$ ns are included in this analysis. The energy of events exceeding the digitizer range was adjusted using Eq. (9.8). The resulting spectrum is shown in the bottom panel of Fig. 9.7.

No peak is apparent at the location predicted by the ^{252}Cf calibration for the inelastic scattering ^{127}I(n,n'γ) reaction. To constrain the total number of inelastic events the following fit procedure was used. First a linearly decaying background was fitted to the full energy range, shown in dashed black. Second, the *ad hoc* peak template, which is shown in Eq. (9.10), was fitted to the peak region. However, most of the fit parameters of the *ad hoc* template were predetermined. The parameters m and c are taken from the fit of the linearly decaying background. N_0, N_1, and δ_N were taken directly from the californium calibration (Table 9.3). The only remaining free parameter was given by the peak amplitude a.

Once the template was fitted to the data the corresponding I was integrated to yield the total number of inelastic scatter events. It is

$$N_{is}^{exp} = 3.9 \pm 11.1,$$

$$(9.16)$$

which is compatible with zero. The 90% confidence level of the upper limit is given by 22.2 counts and its corresponding fit is shown in red in the bottom panel of Fig. 9.7. The fit uncertainty was mainly driven by the overall background achieved and the total exposure recorded. This uncertainty will diminish with more statistics in the future. An MCNPX-PoliMi ver. 2.0 simulation of prompt neutrons was conducted. The spectral hardness and a total flux of these prompt neutrons were set to the values measured in Chap. 4. The neutrons coming from the SNS target unidirectionally bathed the CsI[Na] setup. The total number of inelastic scattering events induced by these prompt neutrons is

$$N_{is}^{sim} = 1.2 \pm 0.2, \tag{9.17}$$

for the full $\mathcal{ON}$ data set, i.e., a total beam energy on target of 7.475 GWh. This prediction is compatible with the value found in Eq. (9.16) at the 1σ level.

The excellent agreement between simulated predictions and measurements confirms the accuracy of the simulation results presented in Chap. 4.

9.4   CEνNS Analysis

In Sect. 9.2 the CsI[Na] detector stability over the course of the 2 year data-taking period was examined. Time periods were identified and excluded from the analysis for which no valid information is available regarding the beam energy delivered on the SNS target. Rare time intervals showing a significant decrease in SPE charge were also excluded. This section provides a detailed description of the post-processing of the n-tuples calculated for the remaining time periods.

As discussed in Chap. 7 the analysis focuses on the residual $\mathcal{R}$ between the $\mathcal{AC}$ and the $\mathcal{C}$ data sets. The former precedes the POT trigger and as such cannot contain any beam-related contributions. It only contains steady-state backgrounds due to environmental radiation. $\mathcal{C}$ in contrast contains both beam-related events and random coincidences. By calculating the residual spectra in energy and arrival time ($\mathcal{R} = \mathcal{C} - \mathcal{AC}$), all contributions from steady-state backgrounds are removed. As a result, $\mathcal{R}$ only contains contributions from beam-related events, be those CEνNS, prompt neutrons, or NINs (Chap. 4).

The theoretical CEνNS prediction can be compared to the residual spectrum $\mathcal{R}$ to determine the observed level of agreement. However, both $\mathcal{R}$ and the CEνNS signal prediction depend on the exact choice of cut parameters used in the post-processing. The following section therefore discusses the analysis used to determine optimized cut parameters, which yield the highest ratio of the expected CEνNS signal to the steady-state environmental backgrounds.

9.4.1   Optimizing Cut Parameters

All data cuts used in the analysis can be classified into three main categories. The first category consists of the quality cuts (Sect. 9.2.4), which cannot be adjusted. The three cuts in this category are the overflow, linear gate, and muon veto cut. These cuts reject a fixed percentage of events and contribute an energy independent signal acceptance fraction. The second category consists of cuts for which the CEνNS search data was used in order to determine their signal acceptance. The first of these is the afterglow cut (Sect. 9.2.3), the second is the diagonal rise-time cut (its definition is analog to that in Sect. 7.4.3), both of which also provide an energy independent signal acceptance fraction. The third category contains cuts for which the ^{133}Ba calibration data was used to calculate a proper signal acceptance (Chap. 7). This category includes the Cherenkov cut (Sect. 7.4.1), as well as the different orthogonal rise-time cuts (Sect. 7.4.2). The signal acceptance obtained from these cuts is given by an energy-dependent acceptance fraction, as described by Eq. (7.8).

In order to establish the optimal choice of cut parameters, a figure of merit (FOM) was defined which quantifies how well a certain set of parameters performs in maximizing the CEνNS signal to steady-state background ratio. In order to prevent introducing any bias, the following approach was chosen: First, the quantification was limited to the $\mathcal{OFF}$ data set, i.e., periods for which there cannot be any beam-related events, neither in $\mathcal{AC}$ nor $\mathcal{C}$. This represents a form of a blind analysis. All data cuts were applied to the $\mathcal{AC}$ data, and the corresponding energy and arrival time spectra of the stead-state background events were determined. Only events with an arrival time of $T_{\text{arr}} \in [0, 6]\mu s$ were included in the energy projection. The arrival time projection only included events with an energy of $N_{\text{pe}} \in [0, 40]$. These choices are based on the expected CEνNS energy and arrival time distributions, which show a negligible CEνNS rate outside of these parameter bounds.

A total of 778,634,460 and 1,558,323,928 triggers were recorded for the $\mathcal{OFF}$ and the $\mathcal{ON}$ data sets, respectively. As discussed earlier, the corresponding beam energy on target delivered for the $\mathcal{ON}$ data is given by $E_{\text{beam}} = 7475\,\text{MWh}$. In order to provide a real background estimate for the CEνNS prediction, the exposure between the $\mathcal{OFF}$ and $\mathcal{ON}$ data-taking periods needs to be matched. To this end, both energy and arrival time spectra determined from the $\mathcal{AC}$ data set were multiplied by a factor of 2.00135. An example of the resulting, scaled spectra for different choices of the Cherenkov cut ($N_{\text{iw}}^{\text{min}}$, Sect. 7.4.1) is shown in panel (**a**) and (**b**) of Fig. 9.9. The number of events passing all cuts decreases significantly with more stringent Cherenkov cut choices, i.e., increasing $N_{\text{iw}}^{\text{min}}$. However, it is also apparent that most of the events rejected carry energies below 20 PE, which is consistent with the energy expected from Cherenkov radiation in the PMT window. In contrast, the arrival time distribution shows an overall decrease in event rate, consistent with what is expected from random coincidences. The slight increase towards earlier arrival times is caused by afterglow. Even though a cut based on $N_{\text{pt}}^{\text{max}}$ was applied to the data set, the arrival time of SPE from phosphorescence from high-energy depositions is still slightly biased towards earlier times in the waveform. This will further be discussed in Sect. 9.4.2.

To quantify the steady-state background in the CEνNS search, the fluctuations σ_{bg} around zero expected for the residual $\mathcal{R}$ in the absence of any beam-related signal (CEνNS, prompt neutrons, or NINs) were estimated. As there is no difference in the steady-state background rate between $\mathcal{AC}$ and $\mathcal{C}$ for the $\mathcal{OFF}$ data set, these fluctuations can be estimated using

$$\sigma_{\text{bg}}^{i} = \sqrt{2N_{AC}^{i}}, \tag{9.18}$$

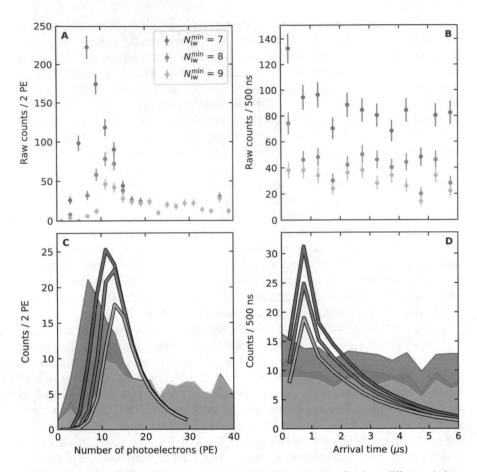

Fig. 9.9 Panel (**a**) and (**b**) show the energy and arrival time spectra for three different choices of the Cherenkov cut N_{iw}^{min}, calculated from $\mathcal{AC}$ of the $\mathcal{OFF}$ data set. The data was scaled to match the number of triggers found in the $\mathcal{ON}$ data set. A significant reduction in the number of events passing the data cuts for increasing N_{iw}^{min} can be observed. Panel (**a**) highlights that mostly events with an energy of less than 20 PE are rejected, whereas panel (**b**) shows an overall reduction independent of the arrival time. This is consistent with the assumption that most of these events consist of random coincidences of Cherenkov spikes with the POT trigger. The trend in time projections towards monotonically decreasing rates with increasing arrival time originates in events with a small afterglow component, able to pass all cuts. The afterglow cut needed to be chosen small enough to reduce this trend to a level at which the $\mathcal{OFF}$ residual shows no systematic deviation from zero in the energy and arrival time spectra. However, softening the afterglow cut increases the number of steady-state backgrounds passing all cuts. This effect is more pronounced in $\mathcal{AC}$ than in $\mathcal{C}$. As a result, the residual spectra show a systematic deviation opposite to a potential CEνNS signal and can therefore not be mistaken for a CEνNS signal. Panels (**c**) and (**d**) show the statistical fluctuations σ_{bg} around zero for the corresponding residual $\mathcal{R}$, caused by random coincidences, as colored bands. The colors correspond to the data shown in panels (**a**) and (**b**). Only the positive fluctuation is shown. The solid lines represent the predicted CEνNS spectra for the different cut values. Comparing the expected CEνNS spectrum to the expected steady-state background allows the definition of a figure of merit (FOM) (Eq. 9.21) which is used to find an optimal set of cut parameters that maximizes the CEνNS signal to steady-state background ratio

where i denotes the ith bin of either energy or arrival time distribution and N^i_{AC} the number of events found with the corresponding energy or arrival time.

The magnitude of these fluctuations is shown in panel (c) and (d) of Fig. 9.9 as colored areas, with only the positive band being shown. It can be observed that the size of the fluctuations naturally decreases with increasing N^{min}_{iw} as fewer and fewer events pass the cuts, i.e., N^i_{AC} becomes smaller and smaller.

To estimate a proper signal-to-background ratio, the expected CEνNS signal for different cut parameter combinations needs to be calculated. For this purpose, the uncut CEνNS-induced nuclear recoil spectrum was calculated as it would be expected in the CsI[Na] detector at the SNS. Combining Eqs. (2.4), (2.3), and (2.5) yields the differential cross-section for both cesium and iodine and with its dependence on the incoming neutrino energy. The differential cross-section was convolved with the three different neutrino emission spectra, Eqs. (3.1), (3.2) and (3.3), to extract a differential recoil spectrum for each neutrino type and isotope. The resulting differential recoil spectrum was further scaled to match a reference beam energy delivered on the SNS target of 1 GWh.

Geant4 simulations regarding the neutrino production rate were conducted by the University of Florida group within the COHERENT collaboration. The full target geometry, the neutron moderators, as well as the beryllium reflector surrounding the target were incorporated in the simulation [7]. The QGSP_BERT physics list was chosen, which uses the Bertini model [2] to simulate the intra-nuclear cascade. A total production rate of $0.08 \pm 10\%$ neutrinos per flavor per proton was found, which amounts to 1.98×10^{21} per GWh per flavor as a reference for a proton energy of ~ 960 MeV.

With a detector mass of 14.57 kg and a distance between target and detector of 19.3 m (Fig. 3.3), the final recoil rate for each flavor and isotope can be determined (Fig. 9.10) for a beam energy delivered on the mercury target of 1 GWh. The total recoil rate is shown in black.

The individual recoil spectra can be integrated to extract the total number of CEνNS events $N^\alpha_{\nu_x}$ that are expected in a zero-threshold detector for each neutrino flavor x and isotope α. The total number of events expected for a perfect detector is given by $N_{total} = \sum_\alpha \sum_x N^\alpha_{\nu_x} = 1146$ events per 1 GWh.

This estimate does not include any threshold effects and other data quality cuts. To properly account for these, the energy deposited in an event needs to be converted into the number of SPE produced. A Monte Carlo approach was used to convert the nuclear recoil energy deposited in the CsI[Na] into a corresponding number of photoelectrons produced N_{pe}, which is described next.

In order to extract a smooth energy spectrum, a total of $N^\alpha_{\nu_x} \times 10,000$ recoil events were simulated for each isotope α and neutrino flavor x. Each event energy was drawn from the corresponding differential recoil spectrum shown in Fig. 9.10. The energy of each event was converted from keV$_{nr}$ to keV$_{ee}$ using the quenching factor calculated in Chap. 8, i.e., $Q = 8.78 \pm 1.66\%$. Using the light yield determined in Chap. 6, i.e., $\mathcal{L}_{CsI} = 13.348 \pm 0.019 \frac{PE}{keV_{ee}}$, the ionization energy was converted to N_{pe}. Poisson fluctuations were applied to the number of photoelectrons. For each

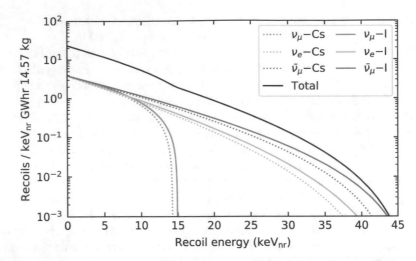

Fig. 9.10 Nuclear recoil rate from CEνNS interactions in the CsI[Na] detector at the SNS. Iodine (cesium) recoils are shown as solid (dotted) lines. The recoil spectra for ^{127}I and ^{133}Cs are very similar as already discussed in the earlier chapters. The total recoil rate is shown in black

event an arrival time was drawn from the distributions shown in the right panel of Fig. 3.1. These results were combined to calculate the uncut CEνNS distributions for all neutrino flavors (Fig. 9.11, panels (**a**) and (**b**)), where the spectra were divided by a factor of 10,000 to account for the MC smoothing factor.

The CEνNS prediction needs to be properly scaled by incorporating the signal acceptance fractions from the different cuts employed in this analysis. An overall signal acceptance function is defined that combines all different cut fractions. This overall acceptance function can be written as

$$\eta(N_{pe}) = \eta_{Ba}(N_{pe}, a, k, x_0)\, \eta_{afterglow}\, \eta_{muon\text{-}veto}\, \eta_{linear\text{-}gate}\, \eta_{diag\text{-}rise\text{-}time} \qquad (9.19)$$

where $\eta_{Ba}(N_{pe}, a, k, x_0)$ is discussed in detail in Chap. 7, whereas $\eta_{afterglow}$, $\eta_{muon\text{-}veto}$, and $\eta_{linear\text{-}gate}$ are discussed in Sect. 9.2. The following approach was used to calculate $\eta_{diag\text{-}rise\text{-}time}$, which was previously motivated and described in Sect. 7.4.3. To determine the acceptance of the diagonal rise-time cut all other cuts had to be applied to the $\mathcal{AC}$ and $\mathcal{C}$ data of the $\mathcal{ON}$ first. Next, the percentage of events with $T_{0-50} \leq T_{10-90}$ was calculated for energies between 50 and 200 PE. As the chance to misidentify the onset of an event is independent of its energy this percentage directly represents the acceptance fraction for the diagonal rise-time cut (Sect. 7.4.3). Figure 9.12 shows the percentage of events satisfying $T_{0-50} \leq T_{10-90}$ as a function of energy for a choice of $N_{pt}^{max} = 3$ and $N_{iw}^{min} = 8$. The dashed blue line represents the uncertainty weighted average of the percentages shown. The 1σ uncertainty is shown as shaded blue region. The acceptance fraction found for this particular choice of afterglow and Cherenkov cut is

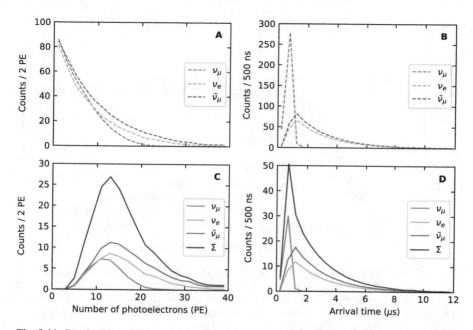

Fig. 9.11 Panels (**a**) and (**b**) show the uncut CEνNS energy and arrival time spectra for each neutrino flavor expected in the CsI[Na] crystal, respectively. The spectra were scaled to match the total neutrino production expected from a beam energy delivered on target of 7475 MWh. Panels (**c**) and (**d**) show the corresponding CEνNS spectra after the signal acceptance function was applied

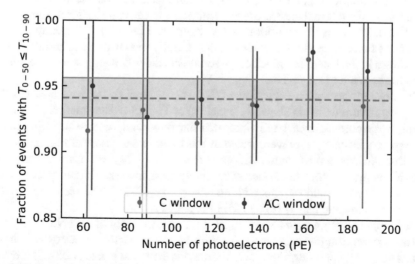

Fig. 9.12 Determination of the signal acceptance fraction for the diagonal rise-time cut using high-energy events in $\mathcal{AC}$ and $\mathcal{C}$. The $\mathcal{AC}$ and $\mathcal{C}$ are offset from their respective bin center to improve readability. The fraction of triggers with a misidentified onset is independent of the energy deposited in an event. As a result one can use high-energy events, which are unaffected by the Cherenkov and the orthogonal rise-time cuts, to determine $\eta_{\text{diag-rise-time}}$

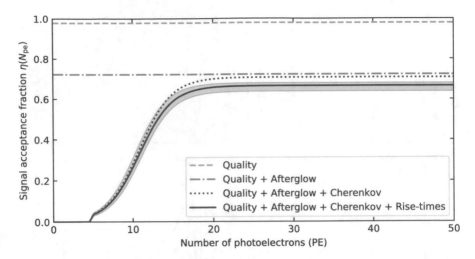

Fig. 9.13 The overall signal acceptance function as calculated for the optimized cut parameters, given in by Eq. (9.23), is shown in black. The associated uncertainty is shown as a gray band and dominated by the available ^{133}Ba statistics that was used to quantify the acceptances of the Cherenkov and orthogonal rise-time cuts. The cumulative evolution of this signal acceptance function upon inclusion of all individual cuts is also shown

$$\eta_{\text{diag-rise-time}} = 0.941 \pm 0.015, \tag{9.20}$$

As an example, Fig. 9.13 shows the overall signal acceptance function corresponding to the optimal choice of cut parameters as calculated later in this section (Eq. 9.23). Shown is the evolution of the acceptance function as different cuts are added. As such, every curve includes all cuts already shown above in addition to the one shown in the label. The black curve represents the full overall signal acceptance function as given in Eq. (9.19).

The overall acceptance function $\eta(N_{\text{pe}})$ was applied to the predicted CEνNS energy spectrum which is shown in panel (**a**) of Fig. 9.11. The fraction of rejected events was recorded for each individual neutrino flavor and the corresponding arrival time spectra for each flavor were scaled by this fraction. For the resulting arrival time spectrum the fraction of events with an arrival time of $T_{\text{arr}} < 6\,\mu\text{s}$ was computed and the energy spectra were scaled accordingly. Last, the arrival time spectrum was adjusted to only include events with an energy of $\leq 30\,\text{PE}$. The resulting CEνNS spectra are shown in panels (**c**) and (**d**) of Fig. 9.11.

It is apparent from panel (**c**) of Fig. 9.11, that a majority of CEνNS events was cut due to overall signal acceptance function. However, for the choice of cuts shown in this example plot (afterglow and rise-time cuts are given in Eq. (9.23), Cherenkov cut $N_{\text{iw}}^{\text{min}}$ as given in the labeling of the figure) one would still expect to observe $\mathcal{O}(100-200)$ events in the energy spectrum.

A simple figure of merit (FOM) can be defined to maximize the CEνNS signal to steady-state background ratio, by demanding that the optimized cut parameters should minimize the background $\sigma_{i,bg}$ while maximizing $N_{i,signal}$ for each bin i at the same time. A straightforward choice for the FOM is to minimize the likelihood of a residual with zero counts and a dispersion of $\sigma_{i,bg}$ to actually mimic a corresponding CEνNS signal. This directly leads to

$$\text{FOM} = \sum_i \frac{N_{i,\,signal}^2}{\sigma_{i,\,bg}^2}, \tag{9.21}$$

which is a χ^2 goodness-of-fit test between the predicted signal and a residual fluctuating around zero counts. In contrast to a maximum likelihood fit, the goal is to minimize the likelihood that a residual fluctuating around zero is able to reproduce the potential CEνNS signal, which means the FOM was maximized.

To identify the optimal cuts, the FOM needs to be calculated for each possible combination of parameters. This cannot be done in a sequential manner, i.e., finding the best parameter choice for one cut, since these cuts interplay on a complex manner. To efficiently find the set of parameters that jointly maximize the FOM, the range of the parameters explored in this search was limited to

$$
\begin{aligned}
\text{Afterglow cut:} \quad & N_{pt}^{max} \in [0; 9] \\
\text{Cherenkov cut:} \quad & N_{iw}^{min} \in [6, 7, 8, 9] \\
\text{Rise-time cuts:} \quad & T_{0-50}^{min} \in [0, 50, 100, 150, 200]\,\text{ns} \\
& T_{0-50}^{max} \in [1.50, 2.00, 2.50, 3.00]\,\mu\text{s} \\
& T_{10-90}^{min} \in [0.00, 0.25, 0.50, 0.75, 1.00]\,\mu\text{s} \\
& T_{10-90}^{max} \in [2.25, 2.50, 2.75, 2.85, 2.95, 3.00]\,\mu\text{s}.
\end{aligned}
\tag{9.22}
$$

This amounts to a total of $10 \times 4 \times 5 \times 4 \times 5 \times 6 = 24{,}000$ potential cut parameter combinations. The rise-time cut parameter ranges were chosen by looking at Fig. 7.11 and estimating an equal percentage of signal events being cut in each step.

For all cut combinations the energy and arrival time spectra were calculated for $\mathcal{AC}$ using the $\mathcal{OFF}$ data set in order to determine $\sigma_{i,\,bg}$ using Eq. (9.18). The acceptance function of the Cherenkov and orthogonal rise-time cuts was calculated using the barium calibration (Chap. 7). By combining all data cuts into a single overall signal acceptance function, (Eq. 9.19), the CEνNS prediction (Fig. 9.11) was properly adjusted. Finally, the FOM was calculated using Eq. (9.21). The parameter choice which maximizes the FOM is given by

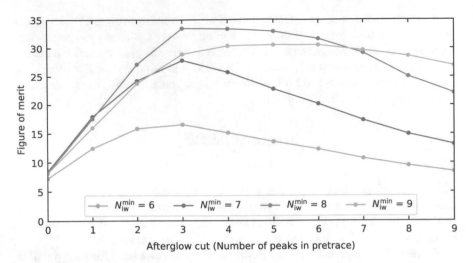

Fig. 9.14 Shown is the FOM calculated using Eq. (9.21) for different afterglow and Cherenkov cut choices. The orthogonal rise-time cuts were fixed to their optimized values as given in Eq. (9.23). It is apparent that $N_{\mathrm{iw}}^{\min} = 8$ yields the highest FOM for most choices of $N_{\mathrm{pt}}^{\max}$

$$
\begin{aligned}
\text{Afterglow cut:} \quad & N_{\mathrm{pt}}^{\max} = 3 \\
\text{Cherenkov cut:} \quad & N_{\mathrm{iw}}^{\min} = 8 \\
\text{Rise-time cuts:} \quad & T_{0-50}^{\min} = 200\,\mathrm{ns} \\
& T_{0-50}^{\max} = 2.5\,\mu\mathrm{s} \\
& T_{10-90}^{\min} = 0.5\,\mu\mathrm{s} \\
& T_{10-90}^{\max} = 2.85\,\mu\mathrm{s}.
\end{aligned}
\tag{9.23}
$$

In order to ensure that this choice of cut parameters indeed represents the best choice and not simply an artifact in the analysis, the evolution of the FOM with respect to the choice of $N_{\mathrm{iw}}^{\min}$ and $N_{\mathrm{pt}}^{\max}$ was investigated. All orthogonal rise-time cuts were set to the values quoted in Eq. (9.23) and the FOM was created for different $N_{\mathrm{pt}}^{\max}$ and $N_{\mathrm{iw}}^{\min}$. The evolution of the FOM is shown in Fig. 9.14. It can be seen that for most choices of afterglow cuts $N_{\mathrm{pt}}^{\max}$ a Cherenkov cut of $N_{\mathrm{iw}}^{\min} = 8$ results in the highest FOM. It can also be verified that the FOM peaks at an afterglow cut of 3 for most Cherenkov cuts. This additional test validates the choice of $N_{\mathrm{pt}}^{\max} = 3$ and $N_{\mathrm{iw}}^{\min} = 8$, which were found in Eq. (9.23).

The overall signal acceptance function associated with this choice of parameters can be calculated using Eq. (9.19). Throughout this thesis these optimal cut parameters were used to illustrate the analysis. By incorporating all individual acceptances Eq. (9.19) can be simplified to

$$\eta_{best}\left(N_{pe}\right) = \frac{0.665}{1 + \dot{e}^{-0.494\,(N_{pe} - 10.85)}}\Theta_{H}\left(N_{pe} - 5\right),\qquad(9.24)$$

where Θ_H again represents the Heaviside step function. This acceptance function is shown in black in Fig. 9.13.

9.4.2 First Observation of CEνNS

Once the cut parameters optimizing the signal-to-background ratio had been calculated, these cuts were applied to all of the data sets, i.e., $\mathcal{AC}$ and $\mathcal{C}$ for both $\mathcal{ON}$ and $\mathcal{OFF}$ data sets. Table 9.4 lists the total number of events passing all cuts for all data set combinations. The large number of events rejected by the Cherenkov and the afterglow cut are apparent. The rise-time cuts only reject a small amount of the surviving events. The numbers presented in Table 9.4 include events of all energies. However, CEνNS-induced recoils only deposit a small amount of energy in the crystal (lower left panel of Fig. 9.11). As such it is favorable to focus on events depositing an energy of $N_{pe} \leq 50\,\mathrm{PE}$ in the crystal. The total number of events passing all cuts in all data sets within this energy region is given by

$$\begin{aligned} N_{\mathcal{AC}}^{\mathcal{OFF}} &= 518 & N_{\mathcal{AC}}^{\mathcal{ON}} &= 1032 \\ N_{\mathcal{C}}^{\mathcal{OFF}} &= 507 & N_{\mathcal{C}}^{\mathcal{ON}} &= 1207. \end{aligned}\qquad(9.25)$$

Table 9.4 Number of events surviving the different cuts applied to the CEνNS search data

Cumulative cuts	Number of waveforms passing all cuts			
Total	2,825,705,648			
Time and power	2,393,035,787			
Quality	2,336,958,388			
	$\mathcal{ON}$		$\mathcal{OFF}$	
	1,558,323,928		778,634,460	
	$\mathcal{C}$	$\mathcal{AC}$	$\mathcal{C}$	$\mathcal{AC}$
Cherenkov	7,298,862	7,362,478	3,320,220	3,353,320
Afterglow	15,779	15,713	7692	7740
Rise-times	12,906	12,844	6228	6270

The cuts are applied in a cumulative manner, as such each new row includes all cuts listed in the rows above

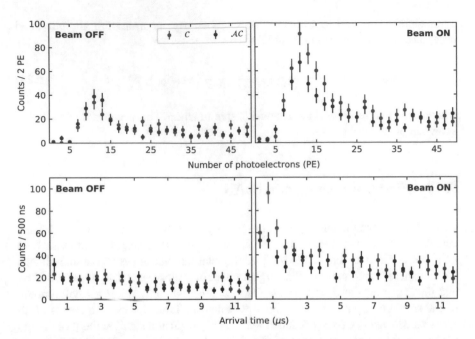

Fig. 9.15 Energy (top) and arrival time (bottom) spectra for all events passing the optimized data cuts. The $\mathcal{OFF}$ ($\mathcal{ON}$) data is shown on the left (right). For times during which the SNS provided beam power on target, i.e., $\mathcal{ON}$, a clear excess of $\mathcal{C}$ over $\mathcal{AC}$ is apparent, both in energy and arrival time. No such excess is visible for $\mathcal{OFF}$

Figure 9.15 shows the energy and arrival time spectra for all events passing the optimized data cuts. Two additional constraints were used during the projection of the data onto the energy and arrival time spectra. First, only events with an arrival time of $T_{arr} \leq 6\,\mu s$ were added to the energy spectrum. Second, the energy of events contributing to the arrival time spectrum was limited to $N_{pe} \leq 30\,PE$, to maximize the contribution from CEνNS-induced events (Panel (**c**) of Fig. 9.9). The $\mathcal{OFF}$ data is shown on the left of Fig. 9.15 and the $\mathcal{ON}$ data set is shown on the right.

In the top left panel of Fig. 9.15 an excess below 15 PE is visible for both $\mathcal{AC}$ and $\mathcal{C}$ in the $\mathcal{OFF}$ data set. This feature is caused by steady-state environmental background in coincidence with the POT trigger. Without data cuts this excess would monotonically increase for $N_{pe} \rightarrow 0$. However, a peak like feature arises once the Cherenkov and rise-time cuts are applied (Fig. 9.13). Both $\mathcal{AC}$ and $\mathcal{C}$ show a comparable number of events which leads to a residual fluctuating around zero. The corresponding arrival time spectrum is shown in the lower left of Fig. 9.15. The slight increase towards earlier arrival times is caused by a small afterglow component from preceding high-energy depositions, introducing a slight bias in the arrival time of the environmental background.

The right panels in Fig. 9.15 show the energy (top) and arrival time spectra (bottom) corresponding to the $\mathcal{ON}$ data set. A clear excess of $\mathcal{C}$ over $\mathcal{AC}$ is apparent in energy and time, demonstrating the presence of a beam-related signal in the data

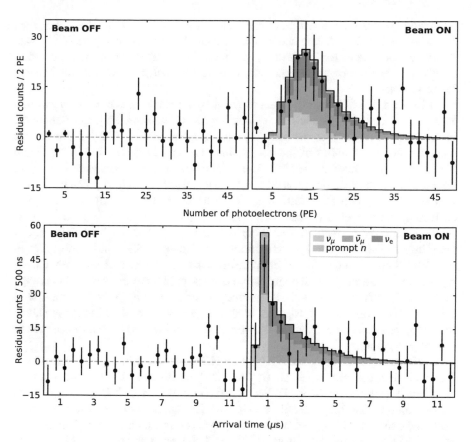

Fig. 9.16 First observation of CEνNS. Residual energy (top) and arrival time (bottom) spectra, i.e., $\mathcal{R} = \mathcal{C} - \mathcal{AC}$, for all events passing the optimized data cuts are shown in black. The $\mathcal{OFF}$ ($\mathcal{ON}$) data is shown on the left (right). Error bars are statistical only. Both energy and arrival time residuals fluctuate around zero for the $\mathcal{OFF}$ data set, confirming that steady-state environmental backgrounds contribute equally to $\mathcal{AC}$ and $\mathcal{C}$. In contrast, the $\mathcal{ON}$ data shows a significant excess both in energy and arrival time. This excess is consistent with the Standard Model CEνNS prediction, shown as stacked green histogram. Every neutrino flavor contributes to the recoil spectrum, as expected from a neutral-current interaction. The presence of a CEνNS signal is favored at 6.7σ over its absence. A small background from prompt neutrons as derived in Chap. 4 is shown in orange. The NIN background was omitted as it consists of ~ 4 counts spread over an energy range of ~ 25 PE and an arrival time range of $\sim 6\,\mu$s. Similar results are obtained from an independent, secondary analysis

set. Comparing the raw number of counts found in $\mathcal{AC}$ for both $\mathcal{OFF}$ and $\mathcal{ON}$ periods showed a comparable level of steady-state environmental background for both data sets. As a result, no degradation in the performance of the optimized cut parameters calculated in Sect. 9.4.1 is to be expected.

The residual $\mathcal{R} = \mathcal{C} - \mathcal{AC}$ was calculated for all spectra shown in Fig. 9.15, which statistically removed all contribution from steady-state backgrounds. The resulting residual spectra are shown in Fig. 9.16 in black, where the error bars are statistical only. The spectra on the left were calculated using $\mathcal{AC}$ and $\mathcal{C}$ of the $\mathcal{OFF}$ data set, whereas the $\mathcal{ON}$ data is shown on the right. The residuals of the $\mathcal{OFF}$ data fluctuate around zero for both energy and arrival time. This was to be expected as steady-state environmental backgrounds contribute equally to $\mathcal{AC}$ and $\mathcal{C}$ and vanish in the subtraction.

In contrast, the $\mathcal{ON}$ residual data shows a significant excess in both energy and arrival time. The $\mathcal{ON}$ panels also include the SM CEνNS prediction as calculated by Grayson Rich [12] as stacked green histograms. Similar CEνNS predictions were found by the author of this thesis, Juan Collar (University of Chicago) and Kate Scholberg (Duke University). The calculations made by Grayson Rich were adopted in this thesis to be consistent with the result published in [7]. The prediction was scaled to match the neutrino emission expected from a total beam energy delivered on the mercury target of 7.475 GWh (Sect. 9.2.1). Rich included most of the secondary corrections that were omitted in Chap. 2, such as axial vector couplings and strange quark contributions [12]. However, the small differences in the CEνNS cross-section for different neutrino flavors due to their different charge radii were neglected. A small contribution from prompt neutrons is shown in orange in Fig. 9.16. This beam-related background was calculated in Chap. 4. Their arrival time is highly concentrated whereas their energy covers most of the range shown in the top panel. The contribution of prompt neutron induced events to each bin is barely discernible.

The beam-related background caused by NINs is omitted in the plot. The reason for this is first, the NIN event rate (Eq. 4.4) is approximately 43 times smaller than the CEνNS event rate (Eq. 9.27). Second, the arrival time of NIN induced events closely follows the 2.2 μs profile and is therefore spread out over the full arrival time range shown. Third, the corresponding events cover a large energy range. As a result the NIN background consists of just $\sim$ 4 counts spread over an energy range of $\sim$ 25 PE and an arrival time range of $\sim$ 6 μs. The NIN contribution to each individual bin is therefore negligible and omitted.

The excellent overall agreement between residual data and the predicted CEνNS signal is apparent. It is further evident that events produced by all three neutrino flavors are necessary to reproduce the experimental data. As CEνNS is a neutral-current process this was to be expected. The full prediction model including CEνNS, prompt neutrons, and NINs is described in more detail in the supplementary online materials of [7].

The total number of events found within a two-dimensional energy-arrival time window bounded by $N_{\mathrm{pe}} \in [6, 30]$ and $T_{\mathrm{arr}} \in [0, 6]$ μs was calculated to be

$$N_{\mathrm{ce}\nu\mathrm{ns}}^{\mathrm{sm}} = 173 \pm 48. \tag{9.26}$$

This corresponds to a CEνNS event rate of

$$\Gamma_{\mathrm{ce}\nu\mathrm{ns}}^{\mathrm{sm}} = 23.1 \pm 6.4 \frac{\text{events}}{\text{GWh}} \tag{9.27}$$

There are mainly four factors contributing to the uncertainty of this prediction. First, the uncertainty on the signal acceptance model contributes $\mathcal{O}(5\%)$. Second, the form factor choice has a significant impact on the overall shape of the nuclear recoil spectrum, which leads to an additional $\mathcal{O}(5\%)$ uncertainty. Third, a $\mathcal{O}(10\%)$ uncertainty is associated with the total neutrino flux emitted by the SNS (Sect. 9.4.1). Last, the quenching factor was calculated in Chap. 8 carrying a $\mathcal{O}(25\%)$ uncertainty. Assuming all of these uncertainties are independent of one another, the total uncertainty on the CEνNS count rate predicted by the Standard Model is $\sim 28\%$.

The simple CEνNS model presented in Sect. 9.4.1 yields the identical number of events within the energy and arrival time boundaries given above. This further illustrates how well the CEνNS cross-section can be approximated by Eq. (2.1) and the relative insignificance of second-order corrections.

In a recent COHERENT publication [7], Rich further performed a binned, maximum likelihood analysis on the two-dimensional energy ($N_{pe} \in [6, 30]$ PE) and arrival time ($T_{arr} \in [0, 6]\,\mu s$) data of the coincidence region ($\mathcal{C}$) in the $\mathcal{ON}$ data set presented in this thesis. The fit model included probability density functions (PDFs) for CEνNS signal, the prompt neutron background, and steady-state environmental backgrounds. No NINs were included in this analysis, due to their negligible contribution. The CEνNS signal PDF was calculated using the same approach as discussed in Sect. 9.4.1, with the inclusion of aforementioned second-order corrections to the CEνNS cross-section. The PDF for prompt neutron induced events was informed by the results presented in Chap. 4. The steady-state environmental background PDF was calculated using the $\mathcal{AC}$ $\mathcal{ON}$ data. Assuming the expected lack of correlation between energy and arrival time for this background, an analytical model was fit to the energy spectrum after the data was marginalized over arrival time. Likewise, the data marginalized over energy was fitted to an analytical model describing the arrival time. The two-dimensional background PDF was constructed by the convolution of both models.

The uncertainties on the prompt neutron rate, the spectral hardness, and quenching factor were included as constraints in the likelihood fit. The amplitude of the steady-state background and its constraint were determined from the $\mathcal{AC}$ data. The amplitude of the CEνNS signal was left unconstrained.

Rich performed this maximum likelihood fit of the $\mathcal{ON}$ residual data derived in this thesis using the RooFit analysis toolkit [15]. The best fit CEνNS signal was found to be [7]

$$N_{ce\nu ns}^{fit} = 134 \pm 22 \qquad (9.28)$$

This value is approximately 23% smaller than the SM prediction, however, it is covered by the 1σ confidence level of $N_{ce\nu ns}^{sm} = 173 \pm 48$.

A more simplistic approach to extract the total number of CEνNS-induced events in the data set is to marginalize over the two-dimensional distributions and to subtract the known background contributions. The total number of events satisfying $N_{pe} \in [6, 30]$ and $T_{arr} \in [0, 6]\,\mu s$ is given by

$$N_{AC}^{OFF} = 209 \qquad\qquad N_{AC}^{ON} = 405$$
$$N_{C}^{OFF} = 209 \qquad\qquad N_{C}^{ON} = 547 \tag{9.29}$$

In Chap. 4 the count rates for the two (prompt neutrons, NINs) main beam-related backgrounds were calculated. Recapitulating these rates were found to be

$$\Gamma_{prompt} = 0.92 \pm 0.23 \, \frac{\text{events}}{\text{GWh}} \tag{9.30}$$

$$\Gamma_{nin} = 0.54 \pm 0.18 \, \frac{\text{events}}{\text{GWh}}. \tag{9.31}$$

Scaling these rates with the total beam energy delivered on target (7.475 GWh) yields a total of 6.9 ± 1.7 counts from prompt neutrons and 4.0 ± 1.3 counts from NINs. As was already discussed in the discussion of beam-related backgrounds (Chap. 4) these backgrounds are ~ 25 (prompt neutrons) and ~ 43 (NINs) times smaller than the SM prediction of CEνNS.

The total number of CEνNS events can therefore be approximated by

$$N_{cevns} = N_{C}^{ON} - N_{AC}^{ON} - N_{prompt} - N_{nin} = 131 \pm 31. \tag{9.32}$$

The total number of CEνNS events determined using this approach agrees well with the number of CEνNS events determined using the profile maximum likelihood fit. N_{cevns} is in agreement with the SM prediction at the 1σ level.

Rich and Barbeau further used the maximum likelihood fit to determine the significance of this observation [7]. The alternative hypothesis, i.e., a CEνNS signal is present, is favored over the null hypothesis, i.e., CEνNS is absent, at the $6.7\text{-}\sigma$ level. This experiment therefore provides ample evidence for the first observation of coherent elastic neutrino-nucleus scattering.

9.4.3 Correlation of Residual Excess and Integrated Beam Energy

To further validate these results the time evolution of the number of residual events $N_{C}^{ON} - N_{AC}^{ON}$ was investigated by the author of this thesis. Figure 9.17 shows the time evolution of the residual counts calculated for both $\mathcal{ON}$ (red) and $\mathcal{OFF}$ (gray) data sets. The accumulated beam energy delivered on target is shown in blue which was normalized to the same vertical scale as the residual axis. The agreement between the increase in residual counts in the $\mathcal{ON}$ data and the total beam energy delivered on the mercury target is evident. In contrast, the $\mathcal{OFF}$ residual only fluctuates around zero. The correlation between beam energy delivered per day and the rate at which $\mathcal{R}$ increases for $\mathcal{ON}$ is also apparent.

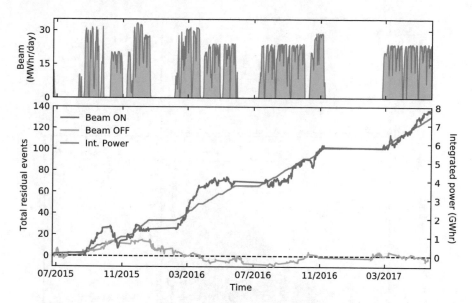

Fig. 9.17 Top: Daily integrated beam power delivered by the SNS on the mercury target, which was discussed in Sect. 9.2.1. **Bottom**: Time evolution of the number of residual counts in both $\mathcal{ON}$ (red) and $\mathcal{OFF}$ (black) data sets. The rate at which the $\mathcal{ON}$ residual grows is correlated to the total integrated beam power acquired (blue). In contrast, the $\mathcal{OFF}$ residual fluctuates around zero. The $\mathcal{OFF}$ residual excess changes continuously due to frequent, short, unplanned outages, not all visible in the top panel

The correlation between beam energy and $\mathcal{ON}$ residual excess was further investigated by Joshua Albert (Indiana University) using a Kolmogorov-Smirnov test [4]. He created several test residual distributions using simulated event distributions. The $\mathcal{ON}$ residual as seen in Fig. 9.17 showed a stronger correlation with the integrated beam power than 96% of all MC generated distributions. It is therefore most likely that the excess found in the $\mathcal{ON}$ residual is produced entirely by beam-related events. The correlation between the residual excess in the $\mathcal{ON}$ data set and the integrated beam energy is unmistakable in Fig. 9.17.

9.4.4 Future Directions for the Analysis Using Statistical Discrimination Between Electronic and Nuclear Recoils

In this section it is investigated whether there is any evidence in the CEνNS search data that nuclear and electronic recoils can be statistically discriminated using the event rise-times [6]. To this end, the rise-time distributions of all events that passed the optimized data cuts were analyzed by the author of this thesis. The distributions for the $\mathcal{OFF}$ data is shown in Fig. 9.18, whereas the $\mathcal{ON}$ data is shown

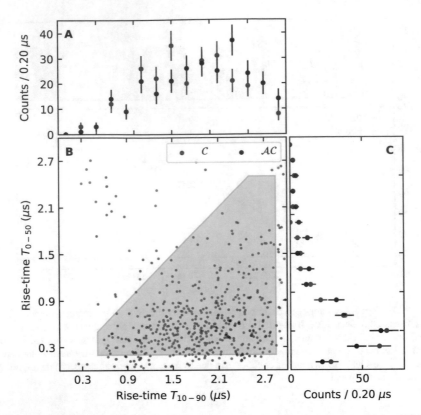

Fig. 9.18 Rise-time distributions of $\mathcal{OFF}$ events passing all optimized data cuts during CEνNS search runs. The $\mathcal{C}$ ($\mathcal{AC}$) data is shown in red (black). The events of both data sets mainly cluster around the same rise-times seen in Fig. 7.11. The rise-time distributions for both $\mathcal{C}$ and $\mathcal{AC}$ look mostly identical. A tail above the $T_{0-50} = T_{10-90}$ diagonal can be identified that is caused by misidentified event onsets due to a preceding SPE. The shaded blue region shows the optimized rise-time cut window

in Fig. 9.19. The $\mathcal{C}$ ($\mathcal{AC}$) data is shown in red (black). The shaded blue region shows the acceptance window of the optimized rise-time cuts used in the CEνNS search (Sect. 9.4.1).

The rise-time distributions for both $\mathcal{OFF}$ and $\mathcal{ON}$ look similar to those extracted from the ^{133}Ba calibration data (Chap. 7, Fig. 7.11). An excess in $\mathcal{C}$ over $\mathcal{AC}$ for the $\mathcal{ON}$ data set can be identified, which is caused by CEνNS-induced nuclear recoils.

To investigate the rise-time distributions of the CEνNS-induced events only, the residual spectra for T_{0-50} and T_{10-90} were calculated. These residual spectra only contain contributions from nuclear recoils. The resulting rise-time distributions for the $\mathcal{ON}$ data set are shown in Fig. 9.20. A broad excess in T_{10-90}, as well as a narrow excess in T_{0-50}, is evident.

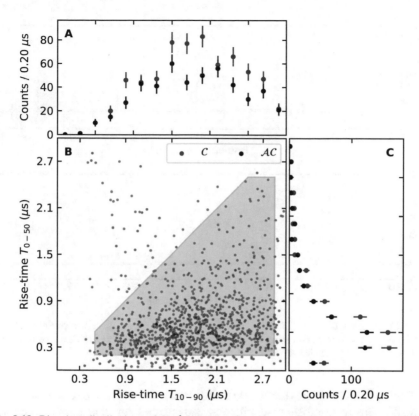

Fig. 9.19 Rise-time distributions of $\mathcal{ON}$ events passing all optimized data cuts during CEνNS search runs. The $\mathcal{C}$ ($\mathcal{AC}$) data is shown in red (black). The events of both data sets mainly cluster around the same rise-times seen in Fig. 7.11. In addition an excess is readily visible for the $\mathcal{C}$ data over the $\mathcal{AC}$ data, which is caused by CEνNS-induced events. This excess is therefore fully comprised of nuclear recoils. The residual $\mathcal{R} = \mathcal{C} - \mathcal{AC}$ is shown in Fig. 9.20. A tail above the $T_{0-50} = T_{10-90}$ diagonal can be identified that is caused by misidentified event onsets due to a preceding SPE. The shaded blue region shows the optimized rise-time cut window

The rise-time residuals for the $\mathcal{OFF}$ data show no significant deviation from zero. The average rise-times for CEνNS-induced nuclear recoils were calculated from the residual spectra. They are

$$\langle T_{10-90} \rangle = (1.84 \pm 0.10)\,\mu s$$
$$\langle T_{0-50} \rangle = (0.68 \pm 0.04)\,\mu s,$$

(9.33)

where the error quoted is the standard error of the mean.

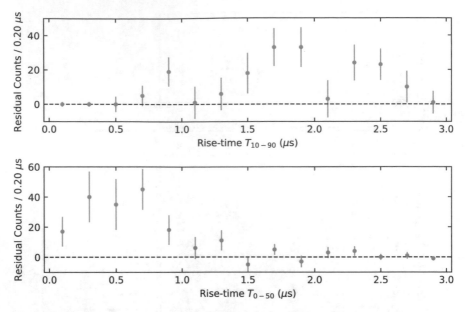

Fig. 9.20 Residual $\mathcal{R} = \mathcal{C} - \mathcal{AC}$ rise-time distributions of all events shown in Fig. 9.19. The distributions show average rise-times differing from those calculated for electronic recoils, hinting at a possible nuclear and electronic recoil discrimination on a statistical basis

The expected rise-times for a nuclear and electronic recoil can be calculated using Eq. (7.5) with fast and slow scintillation decay times for nuclear and electronic recoils as measured in [6]. The rise-times from nuclear and electronic recoils according to Eq. (7.5) are

Nuclear recoils	Electronic recoils	
$T_{10-90} = 1.94^{+0.02}_{-0.06}\,\mu s$	$T_{10-90} = 1.88^{+0.03}_{-0.05}\,\mu s$	(9.34)
$T_{0-50} = 0.56^{+0.01}_{-0.02}\,\mu s$	$T_{0-50} = 0.51^{+0.01}_{-0.01}\,\mu s$	

Comparing the rise-times calculated from the residual data with the rise-times calculated using Eq. (7.5) shows that the current level of statistics is insufficient to discriminate nuclear and electronic recoils. However, the rise-time T_{0-50} measured for the residual spectrum hints at a possible nuclear-electronic recoil discrimination given larger statistics. The CsI[Na] detector is scheduled to continue acquiring data at the SNS. Consequently the statistics will improve and this might provide an opportunity to discriminate nuclear and electronic recoils on a statistical basis.

References

1. J. Amaré, S. Cebrián, C. Cuesta et al., Cosmogenic radionuclide production in NaI(Tl) crystals. J. Cosmol. Astropart. Phys. **2015**(02), 046 (2015), http://stacks.iop.org/1475-7516/2015/i=02/a=046

2. H.W. Bertini, Intranuclear-cascade calculation of the secondary nucleon spectra from nucleon-nucleus interactions in the energy range 340 to 2900 MeV and comparisons with experiment. Phys. Rev. **188**, 1711–1730 (1969), https://link.aps.org/doi/10.1103/PhysRev.188.1711

3. A. Carlson, V. Pronyaev, D. Smith et al., International evaluation of neutron cross section standards. Nucl. Data Sheets **110**(12), 3215–3324 (2009), http://www.sciencedirect.com/science/article/pii/S0090375209001008. Special issue on nuclear reaction data

4. I.M. Chakravarti, R.G. Laha, *Handbook of Methods of Applied Statistics* (Wiley, New York, 1967)

5. S. Chu, L. Ekström, R. Firestone, *The Lund/LBNL Nuclear Data Search. Database Version 2* (1999), pp. 28–99, http://nucleardata.nuclear.lu.se/toi/

6. J.I. Collar, N. Fields, M. Hai et al., Coherent neutrino-nucleus scattering detection with a CsI[Na] scintillator at the SNS spallation source. Nucl. Instrum. Methods Phys. Res. Sect. A Accelerators Spectrom. Detect. Assoc. Equip. **773**, 56–65 (2015), http://www.sciencedirect.com/science/article/pii/S0168900214013254

7. J.I. Collar, D. Akimov, J.B. Albert et al., Observation of coherent elastic neutrino-nucleus scattering. Science **357**(6356), 1123–1126 (2017), http://science.sciencemag.org/content/357/6356/1123

8. N.E. Fields, CosI: development of a low threshold detector for the observation of coherent elastic neutrino-nucleus scattering, Ph.D. thesis (2014), https://search.proquest.com/docview/1647745354?accountid=14657

9. N. Jovančević, M. Krmar, D. Mrda et al., Neutron induced background gamma activity in low-level Ge-spectroscopy systems. Nucl. Instrum. Methods Phys. Res. Sect. A Accelerators Spectrom. Detect. Assoc. Equip. **612**(2), 303–308 (2010), http://www.sciencedirect.com/science/article/pii/S0168900209019640

10. H.S. Lee, H. Bhang, J. Choi et al., First limit on WIMP cross section with low background CsI (Tl) crystal detector. Phys. Lett. B **633**(2–3), 201–208 (2006)

11. T. Nakagawa, K. Shibata, S. Chiba et al., Japanese evaluated nuclear data library version 3 revision-2: JENDL-3.2. J. Nucl. Sci. Technol. **32**(12), 1259–1271 (1995)

12. G. Rich, Measurement of low-energy nuclear-recoil quenching factors in CsI[Na] and statistical analysis of the first observation of coherent, elastic neutrino-nucleus scattering, Ph.D. thesis (2017)

13. B. Scholz, CsI[Na] CEνNS analysis software package at GitHub (2017), https://doi.org/10.5281/zenodo.918287

14. A.B. Smith, P.R. Fields, J.H. Roberts, Spontaneous fission neutron spectrum of ^{252}Cf. Phys. Rev. **108**, 411–413 (1957), https://link.aps.org/doi/10.1103/PhysRev.108.411

15. W. Verkerke, D. Kirkby et al., The RooFit toolkit for data modeling, in *Statistical Problems in Particle Physics, Astrophysics and Cosmology*, ed. by L. Lyons, M.K. Ünel (Imperial College Press, London, 2003), pp. 186–190

Chapter 10
Conclusion

Even though the Standard Model (SM) cross-section predicted for coherent elastic neutrino-nucleus scattering (CEνNS) is the largest of all low-energy neutrino couplings, it had eluded detection for over four decades. This thesis described the efforts of the COHERENT collaboration that resulted in the first observation of CEνNS at a $6.7 - \sigma$ confidence level, using a low-background CsI[Na] detector with a total mass of 14.57 kg. The detector was located at the SNS, a stopped pion source at ORNL, where a pulsed proton beam impinges on a liquid mercury target producing MeV neutrinos. The CsI[Na] was deployed in a sub-basement approximately 20 m away from the target. The location provides $\sim$19 m of continuous shielding against beam-related backgrounds and a total of 8 m.w.e. overburden against cosmic-ray induced backgrounds.

Measurements of the beam-related background at the exact detector location found a background rate from prompt neutrons $\sim$25 times smaller than the expected CEνNS signal. The background caused by NINs originating in the lead surrounding the CsI[Na] detector was calculated to be $\sim$43 times smaller than the expected CEνNS rate. These low-background rates allowed a CEνNS search with a high signal-to-background level.

The CsI[Na] crystal and PMT assembly showed a large light yield of $\mathcal{L}_{csi} = 13.35\frac{\text{SPE}}{\text{keV}_{\text{ee}}}$, which is sufficient to achieve a low-energy threshold. The light collection efficiency was further found to be uniform throughout the whole crystal, greatly simplifying the overall analysis.

To calibrate the analysis pipeline a library of low-energy radiation induced events with a total energy below a few tens of PE was recorded using a ^{133}Ba source. Given the small difference in the scintillation decay times of nuclear and electronic recoils in CsI[Na], this library provides close replicas of the detector response to CEνNS-induced nuclear recoils. It enables calculating the signal acceptance fraction for several data cuts employed in the CEνNS search. The purpose of these data cuts is to reject certain spurious backgrounds, such as Cherenkov light emission in the PMT window, or random groupings of dark-current photoelectrons.

© Springer Nature Switzerland AG 2018

B. Scholz, *First Observation of Coherent Elastic Neutrino-Nucleus Scattering*,
Springer Theses, https://doi.org/10.1007/978-3-319-99747-6_10

In order to calculate the expected CEνNS signal rate in the CsI[Na] detector, the total energy carried by a CEνNS-induced nuclear recoil has to be properly converted into an electron equivalent energy. This conversion makes use of the quenching factor. To reduce the uncertainty associated with the quenching factor, additional measurements were performed at TUNL. These measurements focused on nuclear recoils with an energy of a few keV_{nr} to a few tens of keV_{nr}, covering the energy region expected for CEνNS-induced nuclear recoils at the SNS.

The CsI[Na] detector was deployed at the SNS in June 2015. Since then data was acquired almost continuously at a 60 Hz acquisition rate. The CEνNS search described in this thesis includes all data acquired between June 25th, 2015 and May 26th, 2017. Detector stability measurements show an excellent performance of the setup over the full 2 years of data taking, with only minor exceptions. Using data from time periods during which the SNS did not produce any neutrinos, an optimized set of data cuts was calculated that maximizes the CEνNS signal-to-background ratio. Excesses in energy and event arrival time are observed for neutrino production periods only, which are in agreement with signatures predicted by the SM for CEνNS. The presence of a CEνNS signal is favored at a $6.7 - \sigma$ confidence level over the null hypothesis of environmental background only. The observed CEνNS rate is further compatible with the SM CEνNS cross-section at the $1 - \sigma$ level. The signal rate was further found to be closely correlated to the integrated beam energy delivered on the mercury target, which itself determines the total number of neutrinos produced. This indicates that the measured excess is strictly beam-related.

This observation confirms the coherent enhancement of the neutrino–nucleus scattering cross-section in cesium and iodine at low momentum transfers and substantiates an analogous enhancement in WIMP-nucleus scattering. It also provides confidence in the accuracy of CEνNS cross-sections used in neutrino floor calculations for WIMP searches as well as supernova calculations.

The CsI[Na] detector remains at the SNS and continues to acquire data. The SNS is expected to increase its power output in the near future. During August 2017 the SNS already operated at ~ 1.2 MW resulting in an increase in neutrino production of $\sim 20\%$ compared to earlier months of 2017. A higher beam power is beneficial as it directly increases the CEνNS rate in the CsI[Na], whereas the environmental background rate remains the same. As such the expected signal-to-background ratio also increases with beam power. Combining the analysis presented in this thesis with new CsI[Na] data will therefore result in a CEνNS cross-section measurement with much lower uncertainties.

The COHERENT collaboration aims to measure CEνNS for multiple targets. In addition to the CsI[Na], the collaboration currently operates a 28 kg single-phase LAr detector as well as 185 kg of NaI[Tl]. A planned expansion includes a 2 ton NaI[Tl] array, a ~ 1 ton LAr detector, and ~ 20 kg of p-type point contact germanium detectors. The precision measurements expected from these detectors enable the COHERENT collaboration to pursue new neutrino physics opportunities in the future.